SpringerBriefs in Physics

More information about this series at http://www.springer.com/series/8902

C.J.A.P. Martins

Defect Evolution in Cosmology and Condensed Matter

Quantitative Analysis
with the Velocity-Dependent
One-Scale Model

 Springer

C.J.A.P. Martins
Centro de Astrofísica da Universidade do
    Porto
Porto
Portugal

ISSN 2191-5423   ISSN 2191-5431   (electronic)
SpringerBriefs in Physics
ISBN 978-3-319-44551-9   ISBN 978-3-319-44553-3   (eBook)
DOI 10.1007/978-3-319-44553-3

Library of Congress Control Number: 2016947941

Printed on acid-free paper

This Springer imprint is published by Springer Nature
The registered company is Springer International Publishing AG Switzerland

Preface

The velocity-dependent one-scale (VOS) model was born in the first year of my Ph.D., in early 1995, building upon a previous model by Tom Kibble. In the intervening years it became the canonical model for quantitative studies of the evolution of defect networks, and underwent various important extensions. The purpose of this book is to present a brief overview of the current state of the model.

A literature search will easily reveal several hundred papers (only a fraction of which are my own) discussing many aspects of the VOS model, or directly using it for various purposes. Reviewing, or even citing, all this literature would require a much larger volume and I will not attempt to do so here. Instead, my goal is to provide a unique entry point into the field, by discussing the basic results of the model which the interested reader—perhaps a Masters or Ph.D. student—can learn in a few weeks and use as a starting point for his or her further endeavors.

I thank my Ph.D. supervisor, Paul Shellard, for introducing me to a topic that I still find exciting and challenging. I also thank the rest of my collaborators and my students, for countless interesting discussions on the topic. A large fraction of the numerical work necessary to calibrate this model would not have been possible without the various generations of the COSMOS Shared Memory system at DAMTP, University of Cambridge. At the time of writing, this equipment is operated on behalf of the STFC DiRAC HPC Facility and funded by BIS National E-infrastructure capital grant ST/J005673/1 and STFC grants ST/H008586/1, ST/K00333X/1.

I am supported by an FCT Research Professorship (reference IF/00064/2012), funded by FCT/MCTES (Portugal) and POPH/FSE (EC). This work is also an outcome of project PTDC/FIS/111725/2009 (FCT, Portugal), and I also thank the Galileo Galilei Institute for Theoretical Physics for the hospitality and the INFN for partial support during its completion, and Springer's editor (Angela Lahee) for her flexibility with the agreed deadlines.

Porto, Portugal   C.J.A.P. Martins
June 2016

Contents

Acronyms

BBL	Bazeia–Brito–Losano
CL	Confidence Level
CMB	Cosmic Microwave Background
eLISA	Evolved Laser Interferometer Space Antenna
FRW	Friedmann–Robertson–Walker
GUT	Grand Unified Theory
LHC	Large Hadron Collider
LIGO	Laser Interferometer Gravitational-Wave Observatory
PRS	Press–Ryden–Spergel
RMS	Root-Mean-Squared
VOS	Velocity-Dependent One-scale Model

Chapter 1
Introduction to Defects

Abstract I provide an overview of the theoretical and observational motivations underlying the development of analytic models for the evolution of topological defect networks. This is briefly illustrated for the particular case of cosmic strings. I also highlight links with the more challenging but potentially more interesting case of cosmic superstrings.

1.1 Motivation

Topological defects (cosmic strings, monopoles, domain walls or others) necessarily form at phase transitions in the early universe. This inevitability follows from the Kibble Mechanism. Put simply, it just stems from the fact that there is a cosmological horizon (or in other words, that the speed of light is finite). The type of defects that is formed, as well as their other specific characteristics, will depend on the details of the phase transition, but in most cases the defects will be stable and long-lived. In that case they may be present in the recent universe, as fossil relics of the higher-energy physics. Understanding the evolution and consequences of these defect networks is therefore an unavoidable part of any serious attempt to understand the universe as a whole.

In the 1980 and 1990s, cosmic strings (and, to a lesser extent, other defects) were mostly studied as a competitor to inflationary models in providing the seeds for the formation of the cosmic structures we observe [1]. Cosmic microwave background experiments have now shown that the defect contribution for structure formation must be subdominant (current bounds are at the few percent level). Nevertheless, the fact that they unavoidably form in theoretically preferred scenarios (including models with extra dimensions such as brane inflation) means that they remain an active topic of work. For example, one of the Planck 2013 cosmology papers is dedicated to constraining them [2].

The observational evidence for the accelerating universe shows that our canonical models of cosmology and particle physics are at least incomplete (and possibly incorrect) and that there is new physics waiting to be discovered. After a quest of several decades we now know, thanks to the LHC results, that fundamental scalar

© The Author(s) 2016

C.J.A.P. Martins, *Defect Evolution in Cosmology and Condensed Matter*,
SpringerBriefs in Physics, DOI 10.1007/978-3-319-44553-3_1

fields are among Nature's building blocks. A pressing follow-up question is whether this Higgs-Kibble field has a cosmological role or, more generally, if there are further cosmological scalar fields. Cosmic defects are one of the possible manifestations of such fields. Thus nowadays the search for cosmic defects is part of a wider search for this new physics. Currently the cosmic microwave background provides the most robust constraints, but in the coming years other probes (including ground and space-based gravitational wave detectors, pulsar timing arrays and possibly high-resolution spectroscopy) should provide much tighter constraints.

Quite apart from this specific motivation, the study of their formation and evolution is interesting since it can span an extremely wide range of energy, length and time scales, and yet they are always described by the same underlying physical processes. When studying defect evolution in the early universe one is looking at microscopically small defects, which can nevertheless play an important role on cosmological scales. Examples of other contexts where defects have been studied (and seen) include superconductors, superfluids, liquid crystals, and even tafoni, which can sometimes be seen on the seaside and have also been identified on Mars.

However, in order to make robust predictions of the behavior of these objects that may be accurately compared with the ever-improving observational datasets, one must have detailed models of their evolution. As they are highly non-linear objects, there are two approaches to this: 3D field theory numerical simulations (alternatively, for cosmic strings, 1D Goto-Nambu simulations are also used) or analytic models. The velocity-dependent one-scale (VOS) model is one example of the latter. I developed this for cosmic strings 20 years ago (during my PhD) [3, 4] and it has been successively extended for various other defects and contexts. It remains today the only fully quantitative model of defect network evolution and it is the canonical model in the field (and as such it has been used, for example, in the Planck papers).

Broadly speaking, the ethos of analytic models of cosmic defect networks is akin to starting from statistical physics and turning it into thermodynamics. This approach was first sketched by Kibble [5], but only formulated in a systematic an fully quantitative way with the emergence of the VOS model. Specifically, one starts out from the knowledge of the microphysics of the defects (given, for the case of strings, by the well-known Goto-Nambu action) an uses this to obtain evolution equations for relevant macroscopic quantities, suitably averaged over the network. The two main macroscopic quantities of interest are a characteristic lengthscale (the inter-defect separation, the typical curvature radius, or the network's correlation length) and an averaged (root-mean-square) velocity.

The price for doing this is that in the averaging one most introduce phenomenological parameters whose values cannot be calculated *ab initio*, but must be found by comparing with numerical simulations—thus providing a calibration for the model.

Since the number of free parameters is small, the model is predictive: once the phenomenological parameters are fixed the model will predict what the outcome of simulations with other initial conditions should be. The VOS model has been extensively tested with the world's highest-resolution field theory defect simulations of cosmic strings an domain walls (and to a lesser extent, also of other defects) and thus far a very good agreement has always been found.

The two macroscopic quantities mentioned above are the bare minimum required in order to have a quantitative defect evolution model. Naturally, models with further dynamical variables can be built, and this has also been done in order to describe defects with additional degrees of freedom. This includes scenarios like defects with charges and/or currents, models with additional spacetime dimensions, models with a hierarchy of different tensions, or hybrid networks with defects of different types. The model complexity will of course increase, but the same general principles and formalism will still apply.

The 'classic' book on cosmic defects is Cosmic Strings and Other Topological Defects [6], which was written before the VOS model was developed. This book aims to complement it by providing a focused summary of the work done on the VOS model so far, explaining its overall physical content and describing the various scenarios for which it has been implemented and the predictions of the model in each such scenario.

1.2 An Example: Cosmic Strings

Let us start with a very brief description of a cosmic string [7]. The simplest field theory model that produces them has a single complex scalar field Φ. Let us assume that the Hamiltonian determining the field dynamics is invariant under an axial symmetry, $\Phi \rightarrow \Phi e^{i\theta}$. For example, take the potential energy

$$\int d^3 x V = \int d^3 x \frac{\lambda}{2} \left(|\Phi|^2 - \eta^2 \right)^2 \tag{1.1}$$

where λ is a dimensionless coupling constant and η is an energy scale related to the temperature of the symmetry breaking transition. This has a set of degenerate ground states, the circle $|\Phi| = \eta$, known as the *vacuum manifold*.

At high temperature the field fluctuations are large enough to make the central peak around $|\Phi| = 0$ irrelevant, and the effective potential is symmetric and has a minimum there. As the temperature falls the energy will eventually be too low to permit fluctuations over the peak, at which point the field will settle towards one of the ground states. This random choice of minimum breaks the original axial symmetry. This is the case, for instance, in superfluid ^{4}He.

When a large system goes through a phase transition like this, each part of it has to make this random choice, which need not be the same everywhere. The minimization of gradient terms in the energy of the system tends to make it evolve towards increasingly more uniform configurations, but causality imposes that this evolution can only happen at a limited rate. As a result one expects many domains, each with an uncorrelated choice of ground state. Where these domains meet there is some probability of forming linear defects—cosmic strings—around which the phase angle varies by 2π (or possibly multiples thereof). This is the Kibble mechanism [8]. Notice that the field vanishes at the string's core, so there is trapped potential energy (as well as gradient energy). These strings are known as *global* strings because the transformation $\Phi \to \Phi e^{i\theta}$ is independent of position.

The next step is to consider charged scalar fields interacting with an electromagnetic field. The best known example of a symmetry–breaking transition of this kind is the condensation of Cooper pairs in a superconductor, that has the effect of making photons massive below the critical temperature (in this case the axial symmetry is of the "local" of "gauge" type). The cosmic strings that result are magnetic flux tubes that do not dissipate because the magnetic field is massive outside the string core.

This type of vortex was first discussed by Abrikosov [9] in the context of type II superconductors. Nielsen and Olesen [10] generalized these ideas to the relativistic quantum field theory models used in particle physics, in particular the Abelian Higgs model which is a relativistic version of the Landau–Ginzburg model of superconductivity, governed by the action

$$S = \int d^4x \left[|\partial_\mu \Phi - iqA_\mu \Phi|^2 - \frac{1}{4} F_{\mu\nu} F^{\mu\nu} - \frac{\lambda}{2} (|\Phi|^2 - \eta^2)^2 \right]. \quad (1.2)$$

Here A_μ is the gauge field and Φ is a complex scalar of charge q. The second term is the usual Maxwell action for the electromagnetic field, $F_{\mu\nu} = \partial_\mu A_\nu - \partial_\nu A_\mu$. The energy per unit length of a straight, static string lying on the z-axis is

$$E = \int d^2x \left[|\partial_x \Phi - iqA_x \Phi|^2 + |\partial_y \Phi - iqA_y \Phi|^2 + \frac{1}{2} B^2 + \frac{\lambda}{2} (|\Phi|^2 - \eta^2)^2 \right] \quad (1.3)$$

where $B = \partial_x A_y - \partial_y A_x$ is the z-component of the magnetic field. Finite energy configurations must have $|\Phi| = \eta$ (the *vacuum manifold* is still a circle) but the phase of Φ is undetermined provided the gradient terms and the magnetic field go to zero fast enough. This condition allows for finite energy solutions $A_t = A_r = A_z = 0$, $\Phi(r, \theta) \sim \eta e^{in\theta}$, $A_\theta(r, \theta) \sim n/(qr)$, as $r \to \infty$, in which the total magnetic flux in the plane perpendicular to the string is quantized,

$$\int d^2x B = \oint \mathbf{A} \cdot \mathbf{dl} = \frac{2\pi n}{q}$$

n is the winding number of the string. If the constants λ and q are such that fluctuations in the scalar field Φ and the gauge field A_μ have equal masses, it is possible to show that the string saturates an inequality of the form

Energy per unit length $\geq$ constant x |magnetic flux|

known as the Bogomolnyi bound [11]. In this case, parallel strings at close range exert no force on each other and there are static multivortex solutions [12]. If the mass of the scalar excitations is lower (higher) than that of the gauge excitations, parallel strings will attract (repel).

More complicated particle physics models—in particular those describing the early universe—involve gauge symmetries that generalize the electromagnetic interaction, mediated by photons, to more complicated interactions such as the electroweak or Grand Unified interactions. The messenger fields that play the role of the photons may be massless in the early universe and become massive following a symmetry-breaking transition, and cosmic strings carry the magnetic flux of these other massive gauge fields (not the electromagnetic field).

From a cosmological point of view, the gauge field has the important effect of making the gradient terms decay exponentially fast away from the string so the energy per unit length of these strings is finite. Abrikosov–Nielsen–Olesen strings have no long-range interactions, so their evolution is dominated by their tension and is well described in the thin string or Goto-Nambu approximation.

Field continuity implies that a string of this kind cannot come to an end: it must form a closed loop or extend to infinity, and it cannot break into segments. For this reason, strings, once formed, are hard to eliminate. In the absence of energy loss mechanisms, the strings would eventually dominate the energy density of the universe. On the other hand, the strings can decay into radiation, they may cross and exchange partners, and they may also cross themselves, forming a closed loop which may shrink and eventually disappear. The outcome of these competing mechanisms is that the network is expected to reach a scale-invariant (or *scaling*) regime, where the network's characteristic length scale is proportional to the size of the horizon.

If a random tangle of strings was formed in the early universe, there would always be some strings longer than the horizon, so a few would remain even today. Because cosmological phase transitions typically happen in the very early universe, cosmic strings contain a lot of trapped energy, and can therefore significantly perturb the matter distribution. To first order there is a single parameter quantifying the effects of strings, their energy per unit length. In the simple relativistic strings, the mass per unit length and the string tension are equal, because of Lorentz invariance under boosts along the direction of the string. Cosmic strings are exceedingly thin, but very massive. Typically, for strings produced around the epoch of grand unification, the mass per unit length would be of order $\mu \sim 10^{21} \, \text{kg m}^{-1}$ and their thickness 10^{-24} m. The gravitational effects of strings are effectively governed by the dimensionless parameter $G\mu$, where G is Newton's constant. For GUT-scale strings, this is 10^{-6}, while for electroweak-scale strings it is 10^{-34}.

1.3   Some Observational Consequences

The spacetime around a straight cosmic string is flat. A string lying along the z-direction has an equation of state $p_z = -\rho$, $p_x = p_y = 0$ and therefore there is no source term in the relativistic version of the Poisson equation for the Newtonian gravitational potential

$$\nabla^2 \phi = 0 . \tag{1.4}$$

Nevertheless, a moving string has dramatic effects on nearby matter or propagating microwave background photons. The spacetime metric about such a straight static string has the simple form [13]

$$ds^2 = dt^2 - dz^2 - dr^2 - r^2 d\theta^2 , \tag{1.5}$$

which looks like Minkowski space in cylindrical coordinates, except for the fact that the azimuthal coordinate θ has a restricted range $0 \leq \theta \leq 2\pi(1 - 4G\mu)$. That is, the spacetime is actually conical with a global deficit angle

$$\alpha = 8\pi G\mu , \tag{1.6}$$

where an angular wedge of width α is removed and the remaining edges identified.

This deficit angle implies that the string acts as a cylindrical gravitational lens, creating double images of sources behind the string (such as distant galaxies), with a typical angular separation $\delta\theta$ of order α and no distortion [14]. A long string would yield a distinctive lensing pattern. We should expect to see an approximately linear array of lensed pairs, each separated in the transverse direction. In each lensing event the two images would be identical and have essentially the same magnitude. This is a very unusual signature, because most ordinary gravitational lenses produce an odd number of images of substantially different magnitudes. A number of string lensing event candidates have been discussed in the past, but no confirmed one is currently known.

The above simple picture is complicated by the fact that cosmic strings are not generally either straight or static. Whenever strings exchange partners kinks are created that straighten out only very slowly, so we expect a lot of small-scale structure on the strings. Viewed from a large scale, the effective tension and energy per unit length will no longer be equal. Since the total length of a wiggly string between two points is greater, it will have a larger effective energy per unit length, U, while the effective tension T, the average longitudinal component of the tension force, is reduced, so $T < \mu < U$. This means that there is a non-zero gravitational acceleration towards the string, proportional to $U - T$. Moreover, the strings acquire large velocities, generally a significant fraction of the speed of light, which introduces further corrections to the deficit angle.

Another effect is the formation of over-dense wakes behind a moving cosmic string [15]. When a string passes between two objects, these are accelerated towards each other to a velocity

$$u_\perp = 4\pi G\mu v, \qquad (1.7)$$

where v is the string velocity. Matter therefore collides in a sheet-like structure, leaving a wake behind the moving string. This was the basic mechanism underlying the formation of large-scale structures in cosmic string models, but it fails to reproduce the observed power spectrum of CMB anisotropies; cosmic strings, therefore, can only play a subdominant role in structure formation. Cosmic strings create line-like discontinuities in the cosmic microwave background signal [16, 17]. For the same reason that wakes form behind a cosmic string, the CMB source on the surface of last scattering is boosted towards the observer, so there is a relative CMB temperature shift across a moving string (a red-shift of the radiation ahead of it, and a blue-shift of that behind), given by

$$\frac{\delta T}{T} \sim 8\pi G\mu v_\perp. \qquad (1.8)$$

where $v_\perp$ is the component of the string velocity normal to the plane containing the string and the line of sight. This is known as the Kaiser-Stebbins effect. This simple picture is again complicated in an expanding universe with a wiggly string network and relativistic matter and radiation components. The energy-momentum tensor of the string acts as a source for the metric fluctuations, which create the temperature anisotropies. The most recent comparisons [2] between full-sky maps of cosmic string-induced anisotropies and Planck data yield a cosmological constraint on the models with

$$G\mu < \text{few} \times 10^{-7}, \qquad (1.9)$$

with only a weak dependence on the background cosmology—but a stronger dependence on the modeling of the defects.

Accelerated cosmic strings are sources of gravitational radiation [18]. Consequently, a network of long strings and closed loops produces a stochastic gravitational wave background [19] over a wide range of frequencies and with a spectrum which (at least to a first approximation) has equal power on all logarithmic frequency bins. Another distinctive signal would come from the cusps, the points at which the string instantaneously doubles back on itself, approaching the speed of light. Such an event generates an intense pulse of gravitational and other types of radiation, strongly beamed in the direction of motion of the cusp [20]. If massive cosmic strings do indeed exist, both these pulses and the stochastic background are likely to be among the most prominent signals seen by the gravitational-wave detectors now in operation or planned, in particular Advanced LIGO and eLISA.

A stringent, though indirect, limit on the string energy per unit length comes from observations of the timing of millisecond pulsars. Gravitational waves between us and a pulsar would distort the intervening space-time, and so cause random fluctuations in the pulsar timing. The fact that pulsar timing is extremely regular places an upper limit on the energy density in gravitational waves, and hence on the string scale. The upper limit [21] is nominally of order $G\mu <$ few $\times 10^{-8}$, though there are still considerable uncertainties because this depends on assumptions about the evolution of small-scale structure.

1.4   Cosmic Superstrings

Superstring theory is to date the only candidate model for a consistent quantum theory of gravity that includes all other known interactions. In string theory, the fundamental constituents of nature are not point-like particles but one-dimensional "strings" whose vibrational modes produce all elementary particles and their interactions. Two important features of the theory are supersymmetry (a symmetry between bosonic and fermionic excitations that keeps quantum effects under control) and the presence of extra dimensions above the four spacetime dimensions that we observe.

Before the discovery of D-branes, the "solitons" of superstring theory, the question of whether fundamental superstrings could ever reach cosmological sizes was analysed and the possibility discarded [22]. The discovery of branes and their role in more exotic compactifications where the six compact dimensions have strong gravitational potentials (and redshifts) have changed this picture. It is now believed that networks of cosmic superstrings could be a natural outcome of brane-antibrane annihilation, especially if the branes are responsible for a period of cosmic inflation [23].

An important difference with previous scenarios is that these strings are located in regions of the compactified dimensions with very strong gravitational redshift effects ("warping") that reduce the effective mass per unit length of the strings to a level with deficit angles in the region of 10^{-12}–10^{-7}, compatible with current observations. Another important difference is a much lower probability that the strings intercommute when they cross, estimated to be 10^{-3}–10^{-1}, depending on the type of strings. The lower intercommutation rates lead to much denser networks. Estimates of the corresponding enhancement in the emission of gravitational radiation by cusps puts these strings in a potentially observable window by future gravitational wave detectors [24]. The networks are hybrid, consisting of fundamental strings and D-strings, the latter being either one-dimensional D-branes or perhaps the result of a higher dimensional D-brane where all but one dimension are wrapped around some "holes" (cycles) in the compactified space.

As in the case of hybrid field theory strings, whether or not superstring networks eventually reach a scaling regime is an open question. In addition to the presence of junctions and a non-trivial spectrum of string tensions, a third factor can affect to the evolution of these networks: if the strings are actually higher-dimensional branes partially wrapped around some extra dimensions, then energy and momentum can in principle leak into or out of these extra dimensions [25]. Since the effective damping force affecting the ordinary and extra dimensions is different, one might expect that this will be the case. Depending on its sign and magnitude, such an energy flow can in principle prevent scaling, either by freezing the network (if too much energy leaks out) or by making the strings dominate the universe's energy density (if too much energy leaks in, though this is less likely than the opposite case). In this sense, a somewhat delicate balance may be needed to ensure scaling. At a phenomenological level, further work will be required in order to understand the precise conditions under which each of these scenarios occurs. At a more fundamental level, it is quite likely that which of the scenarios is realized will depend on the underlying compactifications and/or brane inflation models, and that may eventually be used as a discriminating test between string theory realizations.

1.5 Future Directions

A deeper understanding of the evolution and consequences of string networks, especially superstrings, will require progress on both numerical simulations and analytic modelling. At the time of writing there is still no numerical code that includes all the relevant physics, even for the simplest (Goto-Nambu) strings. Nevertheless, domain wall models are being successfully used as toy models.

Inclusion of gravitational backreaction is particularly subtle, and may require completely new approaches. The expected improvements in the available hardware and software will allow for simulations with much longer evolution timespan and spatial resolution, which are needed in order to understand the non-linear interactions between large and small scales all the way down to the level of the constituent quantum fields. This in turn will be a valuable input for more detailed analytic modelling, that must accurately describe the non-trivial small-scale properties of the string networks as well as the detailed features of the loop populations. Better modelling is also needed to describe more general networks—three crucial mechanisms for which at present there is only a fairly simplistic description are the presence of junctions, a non-trivial spectrum of string tensions, and the flow of energy-momentum into extra dimensions.

At a more fundamental level, a better understanding of the energy loss mechanisms and their roles in the evolution of the networks is still missing and it will require new developments in the theory of quantum fields out of equilibrium. Such theoretical developments are also needed to understand defect formation in systems with gauge fields, and could be tested experimentally in superconductors.

The early universe is a unique laboratory, where the fundamental building blocks of nature can be probed under the most extreme conditions, that would otherwise be beyond the reach of any human-made laboratory. Cosmic strings and other defects are particularly interesting for this endeavor: they are effectively living fossils of earlier cosmological phases, where physical conditions may have been completely different. The serendipitous discovery of cosmic defects or other exotic phenomena in forthcoming cosmological surveys will have profound implications for our understanding of cosmological evolution and of the physical processes that drove it. The search continues while, in the meantime, the absence of cosmic string signatures is an increasingly powerful theoretical tool to discriminate between fundamental theories. The possibility that something as fundamental as superstring theory may one day be validated in the sky, using tools as mundane as spectroscopy or photometry, is an opportunity than neither astrophysicists nor particle physicists can afford to miss.

References

1. E.W. Kolb, M.S. Turner, *The Early Universe* (Addison-Wesley, New York, 1994)
2. P.A.R. Ade et al., Astron. Astrophys. **571**, A25 (2014) (Planck Collaboration)
3. C.J.A.P. Martins, E.P.S. Shellard, Phys. Rev. D **53**, 575 (1996)
4. C.J.A.P. Martins, E.P.S. Shellard, Phys. Rev. D **54**, 2535 (1996)
5. T. W. B. Kibble, Nucl. Phys. B **252** (1985) 227; Erratum: [Nucl. Phys. B **261** (1985) 750]
6. A. Vilenkin, E.P.S. Shellard, *Cosmic Strings and other Topological Defects* (Cambridge University Press, Cambridge, 1994)
7. A. Achucarro, C.J.A.P. Martins, Cosmic Strings, in *Encyclopedia of Complexity and Systems Science*, ed. by R. Myers (Springer, New York, 2009)
8. T.W.B. Kibble, J. Phys. A **9**, 1387 (1976)
9. A.A. Abrikosov, Sov. Phys. JETP **5**, 1174 (1957) (Zh. Eksp. Teor. Fiz. **32**, 1442 (1957))
10. H.B. Nielsen, P. Olesen, Nucl. Phys. B **61**, 45 (1973)
11. E.B. Bogomolnyi, Sov. J. Nucl. Phys. **24**, 449 (1976)
12. C.H. Taubes, Commun. Math. Phys. **80**, 343 (1981)
13. A. Vilenkin, Phys. Rev. D **23**, 852 (1981)
14. A. Vilenkin, Astrophys. J. **282**, L51 (1984)
15. J. Silk, A. Vilenkin, Phys. Rev. Lett. **53**, 1700 (1984)
16. N. Kaiser, A. Stebbins, Nature **310**, 391 (1984)
17. J.R. Gott III, Astrophys. J. **288**, 422 (1985)
18. A. Vilenkin, Phys. Lett. B **107**, 47 (1981)
19. C.J. Hogan, M.J. Rees, Nature **311**, 109 (1984)
20. T. Damour, A. Vilenkin, Phys. Rev. Lett. **85**, 3761 (2000)
21. Z. Arzoumanian et al., Astrophys. J. **821**, 13 (2016) (NANOGrav Collaboration)
22. E. Witten, Phys. Lett. B **153**, 243 (1985)
23. S. Sarangi, S.H.H. Tye, Phys. Lett. B **536**, 185 (2002)
24. T. Damour, A. Vilenkin, Phys. Rev. D **71**, 063510 (2005)
25. A. Avgoustidis, E.P.S. Shellard, Phys. Rev. D **71**, 123513 (2005)

Chapter 2
Cosmic Strings

Abstract The velocity-dependent one-scale model of cosmic string network evolution is summarized. Treating the average string velocity as well as the characteristic lengthscale as dynamical variables, one can obtain a fully quantitative model, describing the complete evolution of a string network, including the prediction of previously unknown transient scaling regimes. We also discuss extensions to open, anisotropic and contracting universes, and the effect of radiation backreaction. Finally we discuss the calibration of the model parameters by comparing it to both Abelian-Higgs and Goto–Nambu simulations, in both a static and expanding backgrounds, and highlight the non-trivial fractal properties of cosmic strings.

2.1 Cosmic String Dynamics

We start by summarizing the original derivation of the model discussed in [1, 2]. A string sweeps out a two-dimensional surface (the worldsheet) which can be described by spacetime coordinates x^μ and worldsheet coordinates σ^a, $x^\mu = x^\mu(\sigma^a)$; the line element is then

$$ds^2 = g_{\mu\nu} x^\mu_{,a} x^\nu_{,b} \, d\sigma^a \, d\sigma^b = \gamma_{ab} \, d\sigma^a \, d\sigma^b, \tag{2.1}$$

where $g_{\mu\nu}$ and γ_{ab} are respectively the 4D spacetime and 2D string worldsheet metrics. For the case of a gauge (global) string, one can then derive the Nambu (Kalb-Ramond) action from the Abelian-Higgs (Goldstone) model on the assumption that the scale of perturbations along the string is much larger than its width δ. (In the global case, one also makes use of the equivalence between a real massless scalar field and a two-index antisymmetric tensor field.) One finds

$$S = \begin{cases} \mu_o \int \sqrt{-\gamma} d\sigma^2 & \text{Gauge} \\ \mu_o \int \sqrt{-\gamma} d\sigma^2 + \frac{1}{6} \int \sqrt{-g} H^2 d^4x + 2\pi\eta \int B_{\mu\nu} d\sigma^{\mu\nu} & \text{Global} \end{cases} \tag{2.2}$$

where $B_{\mu\nu}$ is the antisymmetric tensor field, $H_{\mu\nu\lambda}$ is its field strength and $d\sigma^{\mu\nu}$ is the worldsheet area element. Hence the Nambu action is proportional to the area swept out by the string. Varying this action one obtains the equations of motion

© The Author(s) 2016

C.J.A.P. Martins, *Defect Evolution in Cosmology and Condensed Matter*,
SpringerBriefs in Physics, DOI 10.1007/978-3-319-44553-3_2

$$x^\nu_{,a}{}^{;a} + \Gamma^\nu_{\tau\lambda}\gamma^{ab}x^\tau_{,a}x^\lambda_{,b} = \begin{cases} 0 & \text{Gauge} \\ \frac{2\pi\eta}{\mu_o}H^\nu_{\tau\lambda}\,\varepsilon^{ab}x^\tau_{,a}x^\lambda_{,b} & \text{Global} \end{cases} \tag{2.3}$$

Since strings move through a background radiation fluid, their motion is retarded by particle scattering. This effect can be described by a frictional force per unit length [3]

$$\mathbf{F}_f = -\frac{\mu}{\ell_f}\frac{\mathbf{v}}{\sqrt{1-v^2}}, \tag{2.4}$$

where $\mathbf{v}$ is the string velocity and ℓ_f will be called the 'friction lengthscale'; its explicit value depends on the type of symmetry involved. For a gauge string, the main contribution comes from Aharonov–Bohm scattering [4], while in the global case it comes from Everett scattering [5]. Then we have

$$\ell_f = \begin{cases} \frac{\mu}{\beta T^3} & \text{Gauge} \\ \frac{\mu}{\beta T^3}\ln^2(T\delta) & \text{Global} \end{cases} \tag{2.5}$$

where T is the background temperature, δ is the string thickness and β is a numerical factor related to the number of particle species interacting with the string. This force can be included in the equations of motion (2.3) by adding the term

$$\left(U^\nu - x^\nu_{,a}x^{\sigma,a}U_\sigma\right)\frac{1}{\ell_f}, \tag{2.6}$$

(U^ν being the four-velocity of the background fluid) on its right-hand side.

Now consider string motion in an FRW universe with the line element,

$$ds^2 = a^2(\tau)\left(d\tau^2 - \mathbf{dx}^2\right); \tag{2.7}$$

then $U^\nu = \left(a^{-1}, \mathbf{0}\right)$ and choosing the gauge conditions $\sigma^0 = \tau$ (identifying conformal and worldsheet times) and $\dot{\mathbf{x}}\cdot\mathbf{x}' = 0$ (imposing that the string velocity be orthogonal to the string direction) the string equations of motion can be expressed as [3, 6]

$$\ddot{\mathbf{x}} + \left(2\frac{\dot{a}}{a} + \frac{a}{\ell_f}\right)\left(1 - \dot{\mathbf{x}}^2\right)\dot{\mathbf{x}} = \frac{1}{\varepsilon}\left(\frac{\mathbf{x}'}{\varepsilon}\right)', \tag{2.8}$$

$$\dot{\varepsilon} + \left(2\frac{\dot{a}}{a} + \frac{a}{\ell_f}\right)\dot{\mathbf{x}}^2\varepsilon = 0, \tag{2.9}$$

where the 'coordinate energy per unit length' ε is defined by

$$\varepsilon^2 = \frac{\mathbf{x}'^2}{1 - \dot{\mathbf{x}}^2}, \tag{2.10}$$

and dots and primes respectively denote derivatives with respect to τ and σ.

2.1.1 Lengthscale Evolution

We can average the string equations of motion to describe the large-scale evolution of the string network. Define the total string energy and the average RMS string velocity to be

$$E = \mu a(\tau) \int \varepsilon d\sigma, \qquad (2.11)$$

$$v^2 \equiv \langle \dot{\mathbf{x}}^2 \rangle = \frac{\int \dot{\mathbf{x}}^2 \varepsilon d\sigma}{\int \varepsilon d\sigma}. \qquad (2.12)$$

Differentiating (2.11) and using (2.9) and (2.12), we see that the total string energy density $\rho \propto E/a^3$ will obey (in terms of physical time t)

$$\frac{d\rho}{dt} + \left[2H \left(1 + v^2 \right) + \frac{v^2}{\ell_f} \right] \rho = 0. \qquad (2.13)$$

Equation (2.13) incorporates both long strings and small, short-lived loops which usually have a low probability of interacting with other strings before their demise. We shall study the evolution of the long-string network on the assumption that it can be characterized by a single lengthscale L; this can be interpreted as the inter-string distance or the 'correlation length'. Strings larger than L will be called long or 'infinite'; otherwise they will be called loops. For Brownian long strings, we can define the 'correlation length' L in terms of the network density ρ_∞ as

$$\rho_\infty \equiv \frac{\mu}{L^2}. \qquad (2.14)$$

Following Kibble [7], the rate of loop production from long-string collisions can be written as

$$\left(\frac{d\rho_\infty}{dt} \right)_{\text{to loops}} = \rho_\infty \frac{v_\infty}{L} \int w \left(\frac{\ell}{L} \right) \frac{\ell}{L} \frac{d\ell}{L} \equiv \tilde{c} v_\infty \frac{\rho_\infty}{L}, \qquad (2.15)$$

where the loop 'chopping' efficiency $\tilde{c}$ is assumed to be constant. Finally, by subtracting the loop energy losses (2.15) from (2.13) and then using (2.14), we obtain the overall evolution equation for the characteristic lengthscale L,

$$2\frac{dL}{dt} = 2HL(1 + v_\infty^2) + \frac{Lv_\infty^2}{\ell_f} + \tilde{c} v_\infty. \qquad (2.16)$$

Note that with the exception of the expansion term, all terms on the right-hand side are velocity-dependent.

2.1.2 Loop Evolution

Define $n_\ell(\ell, t)d\ell$ to be the number density of loops with length in the range $(\ell, \ell + d\ell)$ at time t; the corresponding loop energy density distribution is

$$\rho_\ell(\ell, t)d\ell = \mu\ell n_\ell(\ell, t)d\ell. \tag{2.17}$$

Note that the total loop energy density is

$$\rho_o \equiv \int \rho_\ell(\ell, t)d\ell; \tag{2.18}$$

the subscript 'o' referring to properties of the entire loop population, while 'ℓ' refers to the loops with length in the range $(\ell, \ell + d\ell)$. From our assumptions on the loop production rate (2.15) we get

$$\frac{d\rho_\ell}{dt} + \left[2H\left(1 + v_\ell^2\right) + \frac{v_\ell^2}{\ell_f}\right]\rho_\ell = g\mu\frac{v_\infty\ell}{L^5}w\left(\frac{\ell}{L}\right), \tag{2.19}$$

where g is a Lorentz factor accounting for the initial non-zero center-of-mass kinetic energy of the loops (lost through velocity redshift). Note that this equation is 'static': it does not include loop decay mechanisms.

The physical size of a loop is simply given by

$$\ell = a(\tau)\int_{loop}\varepsilon d\sigma; \tag{2.20}$$

its time derivative can be easily calculated using (2.9). However one must still subtract energy (hence length) losses due to radiative processes. For a gauge string, this can be roughly estimated from the quadrupole formula

$$\left(\frac{dE}{dt}\right)_{rad} \sim G\left(\frac{d^3D}{dt^3}\right)^2 \sim G\mu^2v^6, \tag{2.21}$$

$(D \sim \mu\ell^3$ being the loop's quadrupole moment). Then we define

$$\left(\frac{d\ell}{dt}\right)_{rad} \equiv -\Gamma'G\mu v^6, \tag{2.22}$$

where according to numerical estimates $\Gamma' \sim 8 \times 65$. Then the evolution equation for the physical loop size is

$$\frac{d\ell}{dt} = (1 - 2v_\ell^2)H\ell - \frac{\ell v_\ell^2}{\ell_f} - \Gamma'G\mu v_\ell^6. \tag{2.23}$$

Now, we will assume that loop production is 'monochromatic', i.e. that loops formed at a time t_p have an initial length

$$\ell(t_p) = \alpha(t_p) \, L(t_p) \,. \tag{2.24}$$

Notice that we are implicitly saying that the loop size at formation depends both on the large-scale properties of the network (through the correlation length) and on the small-scale structure it contains (through the parameter α). With this ansatz the scale-invariant loop production function w becomes

$$w\left(\frac{\ell}{L}\right) = \frac{\tilde{c}}{\alpha} \delta\left(\frac{\ell}{L} - \alpha\right), \tag{2.25}$$

and the rate of energy loss into loops becomes

$$\left(\frac{d\rho_\infty}{dt}\right)_{\text{to loops}} = g\mu\tilde{c}\frac{v_\infty}{L^3} \,. \tag{2.26}$$

Hence the energy density converted into loops from time t to $t + dt$ is

$$d\rho_o(t) = g\mu\tilde{c}\frac{v_\infty}{L^3}dt \,; \tag{2.27}$$

this corresponds to a fraction

$$\frac{d\rho_o(t)}{\rho_\infty(t)} = g\tilde{c}\frac{v_\infty}{L}dt \tag{2.28}$$

of the energy density in the form of long strings at time t. Then using Eq. (2.25), the number of loops produced in a volume V is

$$dN(t) = g\frac{\tilde{c}}{\alpha}\frac{v_\infty}{L^4}V(t)dt \,; \tag{2.29}$$

hence the ratio of the energy densities in 'dynamic' loops and long strings is

$$\varrho(t)_{dyn} \equiv \frac{\rho_o(t)_{dyn}}{\rho_\infty(t)} = gL^2(t)\int_{t_c}^t \frac{dN(t')\ell(t,t')}{V} = g\tilde{c}L^2(t)\int_{t_c}^t \frac{a^3(t')}{a^3(t)}\frac{v_\infty(t')}{L^4(t')}\frac{\ell(t,t')}{\alpha(t')}dt', \tag{2.30}$$

where t_c is the moment of the network formation and $\ell(t, t')$ is the length at time t of loops produced at time t'.

We can also find the ratio of the energy densities in 'primordial' loops and long strings with a modification of our counting strategy: instead of integrating over time, we integrate over the possible loop lengths in the initial distribution

$$\varrho(t)_{pri} \equiv \frac{\rho_o(t)_{pri}}{\rho_\infty(t)} = L^2(t)\frac{a^3(t_c)}{a^3(t)} \int_{L_c}^{L_{cut}} n_\ell(\ell', t_c)\ell(\ell', t_c)d\ell', \tag{2.31}$$

where L_c is the value of the 'correlation length' at time t_c, $L_{cut} \gg L_c$ is a cutoff length, $\ell(\ell', t_c)$ is the length at time t of a (primordial) loop with length ℓ' at t_c and the loop number density n_ℓ has the well-know Vachaspati–Vilenkin form [8]. We can therefore numerically (and, in some simple limit cases, analytically) determine the loop density at all times.

2.1.3 Velocity Evolution

We must now consider the evolution of the average string velocity v. A non-relativistic equation can be easily obtained: it is just Newton's law,

$$\mu\frac{dv}{dt} = \frac{\mu}{R} - \mu v\left(2H + \frac{1}{\ell_f}\right). \tag{2.32}$$

This merely states that curvature accelerates the strings while damping (both from friction and expansion) slows them down. On dimensional grounds, the force per unit length due to curvature should be μ over the curvature radius R. The form of the damping force can be found similarly.

A relativistic generalization can be obtained more rigorously by differentiating (2.12):

$$\frac{dv}{dt} = (1 - v^2)\left[\frac{k}{R} - v\left(2H + \frac{1}{\ell_f}\right)\right]. \tag{2.33}$$

This is exact up to second-order terms. To obtain the damping term we have taken $\langle \dot{\mathbf{x}}^4 \rangle = \langle \dot{\mathbf{x}}^2 \rangle^2$. Writing $\dot{\mathbf{x}}^2 = (1 + \mathbf{p} \cdot \mathbf{q})/2$ ($\mathbf{p}$ and $\mathbf{p}$ being unit left- and right-movers along the string) and defining $\varsigma \equiv -\langle \mathbf{p} \cdot \mathbf{q} \rangle$ the difference between the two is

$$\langle \dot{\mathbf{x}}^4 \rangle - \langle \dot{\mathbf{x}}^2 \rangle^2 = \frac{1}{4}\left[\langle(\mathbf{p} \cdot \mathbf{q})^2\rangle - \varsigma^2\right], \tag{2.34}$$

and numerical simulations of string evolution indicate that $\varsigma_{rad} \sim 0.14$ and $\varsigma_{mat} \sim 0.26$, so this difference should be small [9]. As for the curvature term, we have introduced R via the definition of the curvature radius vector,

$$\frac{a(\tau)}{R}\hat{\mathbf{u}} = \frac{d^2\mathbf{x}}{ds^2}, \tag{2.35}$$

where $\hat{\mathbf{u}}$ is a unit vector and s is the physical length along the string (related to the coordinate length σ by $ds = |\mathbf{x}'|d\sigma = \left(1 - \dot{\mathbf{x}}^2\right)^{1/2} \varepsilon d\sigma$). The dimensionless parameter k is defined by

$$\langle (1 - \dot{\mathbf{x}}^2)(\dot{\mathbf{x}} \cdot \hat{\mathbf{u}}) \rangle \equiv kv(1 - v^2) \tag{2.36}$$

and is related to the presence of small-scale structure on strings: on a perfectly smooth string, $\hat{\mathbf{u}}$ and $\dot{\mathbf{x}}$ will be parallel so $k = 1$ (up to a second-order term as above), but this need not be so for a wiggly string. The following phenomenological function is found to provide a good fit to simulations [10]

$$k(v) = \frac{2\sqrt{2}}{\pi}(1 - v^2)(1 + 2\sqrt{2}v^3)\frac{1 - 8v^6}{1 + 8v^6}. \tag{2.37}$$

If one is only interested in the relativistic regime then

$$k_{\text{rel}}(v) = \frac{2\sqrt{2}}{\pi}\frac{1 - 8v^6}{1 + 8v^6}, \tag{2.38}$$

should be sufficiently accurate to provide reliable results. On the other hand, a reliable approximation for small non-relativistic velocities is

$$k_{\text{nr}}(v) = \frac{2\sqrt{2}}{\pi}(1 - v^2). \tag{2.39}$$

2.2 Scaling Results

In the early universe the friction lengthscale increases with time, so friction will only be important at early times. Let T_c be the temperature of the string-forming phase transition; the corresponding time of formation is

$$t_c = \frac{1}{f}\frac{m_{Pl}}{T_c^2}, \tag{2.40}$$

where $f = 4\pi\sqrt{\pi\mathcal{N}/45}$ and $\mathcal{N}$ is the number of effectively massless degrees of freedom in the model (e.g., $\mathcal{N} = 106.75$ for a minimal GUT model, but it can be as high as 10^4 for particular extensions of it). Then in the case of a gauge symmetry breaking the friction lengthscale can be written

$$\ell_f = \begin{cases} \frac{1}{\theta}\frac{t^{3/2}}{t_c^{1/2}} & \text{Radiation} \\ \left(\frac{3}{4}\right)^{3/2}\frac{1}{\theta}\frac{t^2}{(t_c t_{eq})^{1/2}} & \text{Matter} \end{cases} \tag{2.41}$$

and for the case of a global symmetry

$$
\ell_f = \begin{cases} \frac{1}{4\theta} \frac{t^{3/2}}{t_c^{1/2}} \ln\left(\frac{L}{\delta}\right) \left[\ln\left(\frac{6}{\lambda}\frac{t_c}{t}\right)\right]^2 & \text{Radiation} \\ \left(\frac{3}{4}\right)^{3/2} \frac{1}{4\theta} \frac{t^2}{(t_c t_{eq})^{1/2}} \ln\left(\frac{L}{\delta}\right) \left[\ln\left(\frac{8}{\lambda}\frac{t_c t_{eq}^{1/3}}{t^{4/3}}\right)\right]^2 & \text{Matter} \end{cases}
\tag{2.42}
$$

The constant θ is a measure of the importance of the friction term in the evolution equations; its value is

$$
\theta = \frac{\beta}{\sqrt{f}} \left(\frac{t_c}{t_{Pl}}\right)^{1/2}.
\tag{2.43}
$$

The string energy per unit length can be written

$$
\mu = \begin{cases} T_c^2 & \text{Gauge} \\ T_c^2 \ln\left(\frac{L}{\delta}\right) & \text{Global} \end{cases}
\tag{2.44}
$$

Defining t_* as the time at which the two damping terms in (2.8) and (6.4) have equal magnitude we find

$$
\frac{t_*}{t_c} = \begin{cases} \theta^2 & \text{Gauge} \\ 16\theta^2 \left(\ln\frac{L}{\delta}\right)^{-2} \left[\ln\left(\frac{6}{\lambda}\frac{t_c}{t_*}\right)\right]^{-4} & \text{Global} \end{cases}
\tag{2.45}
$$

provided this is still in the radiation era; otherwise, in the matter era we obtain

$$
\frac{t_*}{t_c} = \begin{cases} \left(\frac{4}{3}\right)^{1/2} \theta \left(\frac{t_{eq}}{t_c}\right)^{1/2} & \text{Gauge} \\ 4\left(\frac{4}{3}\right)^{1/2} \theta \left(\frac{t_{eq}}{t_c}\right)^{1/2} \left(\ln\frac{L}{\delta}\right)^{-1} \left[\ln\left(\frac{8}{\lambda}\left(\frac{t_{eq}}{t_c}\right)^{1/3}\left(\frac{t_c}{t_*}\right)^{4/3}\right)\right]^{-2} & \text{Global} \end{cases}
\tag{2.46}
$$

String dynamics is friction-dominated from t_c until t_*, after which motion becomes relativistic or 'free'. A simple heuristic argument due to Kibble [7] first suggested that in the damped phase the correlation length will scale as $L \propto t^{5/4}$.

Analysis of the evolution equations (2.16), (2.33) reveals the existence of three types of scaling regimes, which we now describe in detail. We should also note that in scaling regimes the velocity and correlation length are generically related via

$$
v \propto \frac{\ell_d}{L},
\tag{2.47}
$$

where we have defined an overall damping length.

$$
\frac{1}{\ell_d} = 2H + \frac{1}{\ell_f}.
\tag{2.48}
$$

2.2.1 Scale-Invariant Solutions

Scale-invariant solutions of the form $L \propto t$ or $L \propto H^{-1}$, together with $v_\infty = const.$, only exist when the scale factor is a power law of the form

$$a(t) \propto t^\lambda, \qquad \lambda = const., \qquad 0 < \lambda < 1 . \qquad (2.49)$$

This condition implies that

$$L \propto t \propto H^{-1} \propto d_H, \qquad (2.50)$$

with the proportionality factors dependent on λ

$$\left(\frac{L}{t}\right)^2 = \frac{k(k + \tilde{c})}{4\lambda(1 - \lambda)}, \qquad v^2 = \frac{k(1 - \lambda)}{\lambda(k + \tilde{c})}, \qquad (2.51)$$

where k is the constant value of $k(v)$ given by solving the second (implicit) equation for the velocity. It is easy to verify numerically that this solution is well-behaved and stable for all realistic parameter values.

2.2.2 Friction-Dominated Solutions

During friction-dominated epochs one has two different scaling solutions, which are transient and no longer 'scale-invariant'. In this case the network retains a memory of its initial conditions, and in particular of the epoch of formation. This can be expressed by the parameter θ in Eq. (2.43) which is the ratio of the damping terms due to friction and Hubble damping, measured at the epoch of string formation.

The first solution is a conformal 'stretching' regime,

$$\frac{L}{L_c} = \left(\frac{t}{t_c}\right)^{1/2}, \qquad v = \frac{t}{\theta L_c}, \qquad (2.52)$$

which will occur when the initial string density and velocity are sufficiently low—for example, as a result of a slow first-order phase transition. In this case the network starts out with a correlation length significantly larger than the damping length and so is 'frozen', and is conformally stretched. However the damping length is growing as $\ell_f \propto t^{3/2}$, so it quickly catches up with it, ending this regime. Although this is not cosmologically relevant except for extremely light strings, the analogous regime in the matter-dominated case would be $L \propto t^{2/3}$, $v \propto t^{4/3}$.

The attractor solution for a friction-dominated epoch, which follows the stretching regime (if this exists) is the Kibble regime, which in the radiation era is

$$\frac{L}{L_c} = \left[\frac{2k_{nr}(\tilde{c} + k_{nr})}{3\theta}\right]^{1/2} \left(\frac{t}{t_c}\right)^{5/4}, \qquad v = \left[\frac{3k_{nr}}{2\theta(\tilde{c} + k_{nr})}\right]^{1/2} \left(\frac{t}{t_c}\right)^{1/4}, \qquad (2.53)$$

where k_{nr} is the value of the momentum parameter in the nonrelativistic limit. In this case the correlation length stays halfway between the damping length and the horizon length. Again there is a matter era analogue, $L \propto t^{3/2}$, $v \propto t^{1/2}$, but this is rarely relevant cosmologically.

2.2.3 A Cosmological Constant

We can also use the VOS model in a flat background to discuss the domination at late times by a cosmological constant. In the extreme asymptotic case when the universe is inflating we have $a \propto \exp(Ht)$ with $H = \sqrt{\Lambda/3}$. The network will 'freeze out' and will simply be conformally stretched, that is,

$$L \propto a, \qquad v_\infty \propto a^{-1}, \qquad (2.54)$$

where, as soon as the strings become nonrelativistic $k_{nr} = 2\sqrt{2}/\pi$, their product satisfies

$$Lv_\infty = \frac{2\sqrt{2}}{\pi}H. \qquad (2.55)$$

2.3 Some Extensions

We now provide short discussions of some extensions of the VOS model. Some of these, as in the case of open universes, mostly have an historic interest, while others are highly relevant. In either case, the goal is to demonstrate the model's versatility, which will be further addressed in the following chapters.

2.3.1 Radiation Back-Reaction

Even though radiation backreaction is closely related to small-scale structure (which the VOS model as described thus far does not explicitly model), its effect on the long-string network can be included in the evolution equation for the correlation length

[10]. For gravitational radiation the following term can be added to the right-hand side of (2.16)

$$2\left(\frac{dL}{dt}\right)_{\text{gr}} \equiv 8\Sigma_{\text{gr}}v_\infty^6 = 8\tilde{\Gamma}G\mu v_\infty^6 \,. \tag{2.56}$$

Here, $\tilde{\Gamma}$ is a constant which is a long-string analogue of the $\Gamma \approx 65$ found for the radiative decay of strings. For global string radiation into Goldstone bosons or axions, the corresponding radiative decay term at a time t will be

$$2\left(\frac{dL}{dt}\right)_{\text{ax}} \equiv 8\Sigma_{\text{ax}}v_\infty^6 = \frac{8\tilde{\Gamma}v_\infty^6}{2\pi \ln(t/\delta)}, \tag{2.57}$$

where the logarithmic term arises because of the long-range fields of the global string and δ is the string width. For GUT-scale strings, the backreaction term for local strings is $\Gamma G\mu \sim 10^{-4}$ whereas for global strings it is of order 10^{-1}.

Remarkably, the inclusion of the back-reaction term does not affect the existence of a scale-invariant attractor solution. However, it does influence the quantitative values of the scaling parameters and the timescale necessary for this solution to be reached: the inclusion of back-reaction can make the approach to scaling much faster.

In this case one can distinguish two asymptotic scenarios. Firstly, if Σ is small (of order unity at most) then the effect of back-reaction on the scaling solution is also small. This will be the case, for example, for most local or global string networks in a cosmological context. We can express this as

$$\gamma^2 \approx \gamma_0^2\,(1+\Delta)\,, \qquad v^2 \approx v_0^2\,(1-\Delta)\,, \tag{2.58}$$

where γ_0 and v_0 are the "unperturbed" scaling values, given by Eq. (2.51), and the back-reaction correction has the form

$$\Delta = 8\beta v_0^5 \Sigma = 8\beta\left[\frac{k(1-\beta)}{\beta(k+\tilde{c})}\right]^{5/2}\Sigma\,. \tag{2.59}$$

Second, for larger values of Σ the back-reaction term will dominate the evolution equation for L, and the attractor scale-invariant solution has a different form altogether. It is not possible to write this solution in closed form, even expressing k implicitly as above. However, it is possible to write it as a series. The dominant term and the first correction take the form

$$\gamma = \frac{k}{2\beta}\left[\frac{8\beta\Sigma}{k(1-\beta)}\right]^{1/7}(1+\Delta_1+\ldots)\,, \qquad v = \left[\frac{k(1-\beta)}{8\beta\Sigma}\right]^{1/7}(1-\Delta_1+\ldots)\,, \tag{2.60}$$

with

$$\Delta_1 = \frac{1}{2^{6/7}7}(k+\tilde{c})\left[\frac{\beta}{k(1-\beta)}\right]^{5/7}\Sigma^{-2/7}\,. \tag{2.61}$$

There has been work on numerical simulations of global string networks [11] which explores this strong backreaction regime. These authors report a surprisingly low string density relative to the gauged case. For their expanding universe simulations in the realistic case with periodic boundary conditions, they find the following radiation and matter era densities respectively,

$$\zeta_{\text{rad}} = 0.9 \pm 0.1\,, \qquad \zeta_{\text{mat}} = 0.5 \pm 0.1\,. \tag{2.62}$$

These results are perfectly consistent (within the estimated error bars) with our extended VOS model if we adopt a back-reaction parameter

$$\Sigma_{\text{ax-sim}} \approx 3\,. \tag{2.63}$$

Indeed, this corresponds to the approximate average value for Σ_{ax} that one would estimate for simulations of this resolution. Present limitations on numerical dynamic range give the upper bound $\ln(t/\delta) < 4$ (at the end of the simulation), implying $\Sigma_{\text{ax-sim}} > 2$ throughout.

2.3.2 Open Universes

In this section we discuss the behavior of our model in open universes, for which one must also include an additional correction due to the curvature [12–14]. This is essentially the curvature radius of the strings, which as previously said we assume to coincide with L, divided by the radius of spatial curvature of the universe,

$$\mathscr{R} = \frac{H^{-1}}{|1 - \Omega|^{1/2}}\,. \tag{2.64}$$

After a certain amount of algebra, one finds correction terms of the form

$$w = 1 - (1 - \Omega)(HL)^2\,. \tag{2.65}$$

Note that Ω denotes the total density of the universe. For a flat universe, $\Omega = 1$, and we have $w = 1$. The evolution equation for the correlation length L now takes the form

$$2\frac{dL}{dt} = 2HL + \frac{L}{\ell_d}\frac{v_\infty^2}{w^2} + \tilde{c}v_\infty\,, \tag{2.66}$$

while the velocity equation becomes

$$\frac{dv_\infty}{dt} = \left(1 - \frac{v_\infty^2}{w^2}\right)\left(w^2\frac{k}{L} - \frac{v_\infty}{\ell_d}\right)\,. \tag{2.67}$$

We can now re-examine the question of the existence of 'scale invariant' attractor solutions. Scaling solutions of the form $L \propto t$ or $L \propto H^{-1}$ together with $v_\infty = const.$ still only exist provided

$$a(t) \propto t^\lambda, \qquad \lambda = const., \qquad 0 < \lambda < 1, \tag{2.68}$$

but now we also require

$$\Omega = const. \tag{2.69}$$

The simplest example of the second condition is of course a flat universe, but there are examples of cosmological models which have attractors other than $\Omega = 1$ [15]. In any case, note that there can be additional relations between the values of λ and Ω for specific models. The scaling solution is now given in the implicit form

$$\left(\frac{L}{t}\right)^2 = w^2 \frac{k(k + \tilde{c})}{4\lambda(1 - \lambda)}, \qquad v^2 = w^2 \frac{k(1 - \lambda)}{\lambda(k + \tilde{c})}, \tag{2.70}$$

where k is defined as before, and

$$w = \frac{2(1 - \lambda)}{(1 - \Omega)\lambda k(k + \tilde{c})} \left[\left(1 + \frac{(1 - \Omega)\lambda k(k + \tilde{c})}{(1 - \lambda)}\right)^{1/2} - 1\right]. \tag{2.71}$$

If the two conditions above do not hold, then a scaling solution will not exist.

We should also mention another cosmologically important solution: in an open universe with $\Omega \to 0$, $a \propto t$, the asymptotic solution is

$$L = \left[\frac{k_{nr}\tilde{c}}{2(1 - k_{nr})}\right]^{1/2} t \, (\ln t)^{1/2}, \qquad v_\infty = \left[\frac{k_{nr}(1 - k_{nr})}{2\tilde{c}}\right]^{1/2} (\ln t)^{-1/2}, \tag{2.72}$$

with k_{nr} given by (2.39). Note that this *is not* a scale-invariant solution, since $H^{-1} = t$ and $d_H = t \ln t$. In other words, by looking at the network one would be able to determine when the curvature-dominated period had started.

2.3.3 Anisotropic Models

Topological defects can be a relic left behind after inflation. The inflationary epoch will dilute the defect density and push the network outside the horizon, freezing it in co-moving coordinates in the process. However, once the inflationary epoch ends the subsequent evolution of the defects is necessarily such as to make them come back inside the horizon [13, 16]. A defect network produced during an anisotropic phase in the very early universe could still be present today.

Let us consider a cosmic string in a flat anisotropic universe of Bianchi type I, with line element:

$$ds^2 = dt^2 - X^2(t)x^2 - Y^2(t)dy^2 - Z^2(t)dz^2 . \tag{2.73}$$

Here $X(t)$, $Y(t)$ and $Z(t)$ are the cosmological expansion factors in the x, y and z directions respectively, and t is physical time. We also define $A \equiv \dot{X}/X$, $B \equiv \dot{Y}/Y$ and $C \equiv \dot{Z}/Z$ where the dot represents a derivative with respect to physical time t.

In the limit where the curvature radius of a cosmic string is much larger than its thickness, we can describe it as a one-dimensional object so that its world history can be represented by a 2D world-sheet

$$x^\nu = x^\nu(\zeta^a); \qquad a = 0, 1; \quad \nu = 0, 1, 2, 3 \tag{2.74}$$

obeying the usual Goto–Nambu action

$$S = -\mu \int \sqrt{-\gamma} d^2\zeta , \tag{2.75}$$

where μ is the string mass per unit length, γ_{ab} is the two-dimensional world-sheet metric and $\gamma = \det(\gamma_{ab})$. Let us also define

$$\dot{\mathbf{x}}^2 \equiv g_{\alpha\beta}\dot{x}^\alpha \dot{x}^\beta = 1 - X^2\dot{x}^2 - Y^2\dot{y}^2 - Z^2\dot{z}^2 \tag{2.76}$$

$$\mathbf{x}'^2 \equiv g_{\alpha\beta}x'^\alpha x'^\beta = -X^2x'^2 - Y^2y'^2 - Z^2z'^2 , \tag{2.77}$$

so that $\gamma = \dot{\mathbf{x}}^2\mathbf{x}'^2$ (using $\dot{\mathbf{x}} \cdot \mathbf{x}' \equiv g_{\alpha\beta}x'^\alpha \dot{x}^\beta = 0$ as a gauge condition).

If we choose $\zeta^0 = t$ and define $\zeta \equiv \zeta^1$ then the string equation of motion is given by [9]

$$\frac{\partial}{\partial t}\left(\frac{\dot{x}^\mu \mathbf{x}'^2}{\sqrt{-\gamma}}\right) + \frac{\partial}{\partial\zeta}\left(\frac{x'^\mu \dot{\mathbf{x}}^2}{\sqrt{-\gamma}}\right) + \frac{1}{\sqrt{-\gamma}}\Gamma^\mu_{\nu\sigma}(\mathbf{x}'^2\dot{x}^\nu \dot{x}^\sigma + \dot{\mathbf{x}}^2 x'^\nu x'^\sigma) = 0 . \tag{2.78}$$

From the time component we can obtain

$$\dot{\varepsilon} + \varepsilon\left[AX^2\left(\dot{x}^2 - \frac{x'^2}{\varepsilon^2}\right) + BY^2\left(\dot{y}^2 - \frac{y'^2}{\varepsilon^2}\right) + CZ^2\left(\dot{z}^2 - \frac{z'^2}{\varepsilon^2}\right)\right] = 0 \tag{2.79}$$

where we have made the further definition

$$\varepsilon \equiv \sqrt{-\mathbf{x}'^2/\dot{\mathbf{x}}^2} = \mathbf{x}'^2/\sqrt{-\gamma} = \sqrt{-\gamma}/\dot{\mathbf{x}}^2 . \tag{2.80}$$

On the other hand, the x component gives

$$\ddot{x} + \left(\frac{\dot{\varepsilon}}{\varepsilon} + 2A\right)\dot{x} + \frac{1}{\varepsilon}\left(\frac{x'}{\varepsilon}\right)' = 0, \tag{2.81}$$

and analogous equations apply for the y and z components. One can show that in the limit of an isotropic universe these equations reduce to the usual form.

Further analysis shows that the existence of an anisotropic phase through which the network evolved will be imprinted on it much beyond the time when the background becomes isotropic [13, 16]. In fact, it will be imprinted on the network as long as it is frozen outside the horizon. Only when it falls inside the horizon it will start to become relativistic and isotropic. We expect that the evolution towards the relativistic regime will be somewhat slower than in the standard case, which could conceivably have observational implications. Specifically, it is possible to see that a cosmic string network can survive up to about 60 e-foldings of inflation (the exact number being model-dependent), in the sense that any network produced in such a period will still come back inside the horizon in time to have observable consequences by the present day.

2.4 Calibrating the Model with Simulations

The model has been calibrated by detailed comparisons of its predictions to Abelian-Higgs (field theory) [17, 18] and Goto–Nambu numerical simulations [10, 19]. In the field theory case the best fit is provided by

$$\tilde{c} = 0.57 \pm 0.04, \tag{2.82}$$

which is is precisely the same value that was found in flat spacetime Goto–Nambu string simulations. On the other hand, for radiation and matter era Goto–Nambu simulations, one finds

$$\tilde{c} = 0.23 \pm 0.04. \tag{2.83}$$

This is to be expected given that Goto–Nambu network simulations can probe a much wider range of length scales below the correlation length, thus allowing small-scale wiggles to build up on those scales. We emphasize that no currently available field theory simulation has a spatial resolution or dynamic range sufficiently large to allow for the build-up of small-scale structures on the strings.

The value of $\tilde{c} = 0.57$ can therefore be regarded as a bare loop chopping efficiency, while $\tilde{c} = 0.23$ can be interpreted as a renormalised one. This interpretation is consistent with the fact that Goto- Nambu simulations in the expanding case somewhat surprisingly possess much more small-scale structure than
corresponding flat spacetime strings (for example, as quantified by the fractal
properties of each network). The approximate factor of two difference between
the two loop production rates may be related to the well-known result that the
renormalised and bare string mass per unit length differ by about a factor of
two in radiation era Nambu simulations [19–21].

The most detailed numerical study of the properties of small-scale structures on
cosmic string networks has been carried out in [19]. A first striking feature is that in
the expanding universe cosmic string velocities are *anti-correlated* on scales between
the correlation length and the horizon. However, such a feature is not present in flat
spacetime. This anti-correlation is the result of a 'memory' of the network for recent
intercommutings, and its absence in the flat spacetime case highlights the fact that
the loop production mechanism is different in the expanding and non-expanding
cases. Indeed, such an effect was discussed in [22, 23]. If one defines a 'velocity
coherence length', this will be significantly smaller than ξ itself. The network's fractal
dimension is unity on small scales and two on very large scales. The interesting
question, however, is what happens at intermediate scales. In particular, one should
expect (and indeed finds) a range of scales where strings should behave as *self-
avoiding random walks*, and these have a fractal dimension $d_s = 5/3$ in three spatial
dimensions.

Radiation and matter era networks have similar fractal profiles (if one rescales
length scales by the respective correlation lengths). The flat spacetime ones, however,
are qualitatively different, which must reflect the differing efficiencies of loop production in flat space and the expanding universe. While the integrated loop production
efficiency is much greater in flat spacetime $c = 0.57$, as opposed to the expanding
$c = 0.23$, it appears to be relatively less effective around the correlation scale with
energy trapped on fairly large scales. This intuitive picture is confirmed by noting
that the renormalised mass per unit length μ is smaller than in the expanding case
on small scales, but is larger on large scales.

Finally, we note that in all cases the fractal dimension of the network at the scale
of the correlation length is well below two: typical values are 1.2 in the expanding
case and 1.4 in the flat case. This is to be expected if one interprets ξ as a persistence
length. So the intuitive picture that a string network looks Brownian at the scale
of the correlation length is clearly incorrect. It's not even true at the scale of the
horizon—here the network looks more like a *self-avoiding* random walk, which is
an obvious consequence of intercommutings. The Brownian picture is only valid on
significantly larger scales.

The two distinguishing characteristics of string evolution in Minkowski spacetime are the absence of velocity anti-correlations on scales around the correlation length, and the apparent existence of a 'preferred' scale (around the correlation length ξ) from which energy does not move to smaller scales. These can have a substantial influence when calculating the network's observational consequences.

References

1. C.J.A.P. Martins, E.P.S. Shellard, Phys. Rev. D **53**, 575 (1996)
2. C.J.A.P. Martins, E.P.S. Shellard, Phys. Rev. D **54**, 2535 (1996)
3. A. Vilenkin, Phys. Rev. D **43**, 1060 (1991)
4. R. Rohm, Ph.D. thesis, Princeton University (1985)
5. A.E. Everett, Phys. Rev. D **24**, 858 (1981)
6. N. Turok, P. Bhattacharjee, Phys. Rev. D **29**, 1557 (1984)
7. T. W. B. Kibble, Nucl. Phys. **B252**, 227 (1985); **B261**, 750 (1986)
8. T. Vachaspati, A. Vilenkin, Phys. Rev. D **30**, 2036 (1984)
9. A. Vilenkin, E.P.S. *Shellard, Cosmic Strings and other Topological Defects* (Cambridge University Press, Cambridge, 1994)
10. C.J.A.P. Martins, E.P.S. Shellard, Phys. Rev. D **65**, 043514 (2002)
11. M. Yamaguchi, Phys. Rev. D **60**, 103511 (1999); M. Yamaguchi, J. Yokoyama, M. Kawasaki, Phys. Rev. **D61**, 061301 (2000)
12. C.J.A.P. Martins, Phys. Rev. D **55**, 5208 (1997)
13. P.P. Avelino, R.R. Caldwell, C.J.A.P. Martins, Phys. Rev. D **59**, 123509 (1999); P.P. Avelino, C.J.A.P. Martins, Phys. Rev. D **62**, 103510 (2000)
14. Martins, C.J.A.P., Quantitative string evolution, Ph.D. thesis, University of Cambridge (1997)
15. J.D. Barrow, J. Magueijo, Class. Quant. Grav. **16**, 1435 (1999). Phys. Lett. **B447**, 246 (1999)
16. J.R.C.C. Correia, I.S.C.R. Leite, C.J.A.P. Martins, Phys. Rev. D **90**, 023521 (2014)
17. J.N. Moore, E.P.S. Shellard, C.J.A.P. Martins, Phys. Rev. D **65**, 023503 (2002)
18. C.J.A.P. Martins, J.N. Moore, E.P.S. Shellard, Phys. Rev. Lett. **92**, 251601 (2004)
19. C.J.A.P. Martins, E.P.S. Shellard, Phys. Rev. D **73**, 043515 (2006)
20. D.P. Bennett, F.R. Bouchet, Phys. Rev. D **41**, 2408 (1990)
21. B. Allen, E.P.S. Shellard, Phys. Rev. Lett. **64**, 119 (1990)
22. E.P.S. Shellard, B. Allen, On the evolution of Cosmic Strings, in *The Formation and Evolution of Cosmic Strings*, ed. by G.W. Gibbons et al. (Cambridge University Press, Cambridge, 1990)
23. D. Austin, E.J. Copeland, T.W.B. Kibble, Phys. Rev. D **48**, 5594 (1993)

Chapter 3
Domain Walls

Abstract The VOS model for cosmic strings is now extended to the case of domain wall networks, and calibrated against high-resolution field theory numerical simulations in two, three and four spatial dimensions. We briefly study domain wall forming models where different tensions and various types of defect junctions can exist, which illustrate some of the mechanisms that will determine the evolution of defect networks with junctions. We find that the networks reach the attractor linear scaling solutions in all such cases, and also provide state-of-the-art constraints on these networks. We then study the evolution of various types of biased domain wall networks, discussing possible mechanisms of decay of these networks. Finally, we revisit the model and present an alternative formulation in terms of a physical (rather than invariant) characteristic length scale, which we use to study the evolution of domain wall and cosmic string networks in contracting universes.

3.1 The VOS Model for Domain Walls

The wall surface $\mathcal{M}_2$ can be parametrized by two parameters, σ_1 and σ_2. As a result, the wall evolution is described by the vector $x^\mu(\sigma_1, \sigma_2, \tau)$, where we identified $\sigma_0 = \tau$. If the function is smooth, it is possible to parametrize the wall surface such that two tangential vectors will be orthogonal

$$\partial_{\sigma_1} x^\mu \partial_{\sigma_2} x_\mu \equiv x^\mu_{,1} x_{\mu,2} = 0. \tag{3.1}$$

We can also require that the velocity of the wall $\partial_\tau x^\mu \equiv \dot{x}^\mu$ is normal to the tangent surface. To derive the wall equation of motion we start from the worldvolume (Dirac) action, which is the extension of the Goto-Nambu one for cosmic strings and has the form

$$S = -\int \mathcal{L} d^3\sigma = -\sigma_w \int \sqrt{\gamma} d^3\sigma, \tag{3.2}$$

© The Author(s) 2016

C.J.A.P. Martins, *Defect Evolution in Cosmology and Condensed Matter*,
SpringerBriefs in Physics, DOI 10.1007/978-3-319-44553-3_3

where σ_w is a constant mass per unit area, $\gamma_{ab} = g_{\mu\nu} x^\mu_{,a} x^\nu_{,b}$ is the induced metric, $\gamma = \frac{1}{3!} \varepsilon^{ab} \varepsilon^{cd} \gamma_{ac} \gamma_{bd}$ is its determinant, $x^\mu_{,a} = \frac{\partial x^\mu}{\partial \sigma^a}$, ε^{ab} is the Levi-Civita symbol, and $\mathscr{L}$ is the Lagrangian density. The energy of the wall in that case is

$$E = \sigma_w a(\tau) \int \sqrt{\gamma} \gamma^{00} d^2\sigma = \sigma_w a^2(\tau) \int \varepsilon d^2\sigma. \tag{3.3}$$

In a flat FRW universe the equation of motion has the form

$$\frac{\dot{a}}{a} \delta_{0\lambda} \sqrt{\gamma} \gamma^{ab} \gamma_{ab} - \partial_c \left(\sqrt{\gamma} \gamma^{ab} g_{\mu\lambda} x^\mu_{,a} \delta^c_b \right) = 0. \tag{3.4}$$

Let us redefine the coordinates σ_1 and σ_2 to s_1 and s_2 in such way that $|\frac{\partial x^i}{\partial s_\alpha}|^2 = 1$ ($\alpha = 1, 2$). This means that derivatives will be changed in the following way

$$\frac{\partial x^i}{\partial \sigma_\alpha} = |x^i_{,\alpha}| \frac{\partial x^i}{\partial s_\alpha}, \tag{3.5}$$

(no summation over α). In these coordinates, it is possible to introduce an orthonormal basis: $\xi^i_\alpha = \frac{\partial x^i}{\partial s_\alpha}$, and $n^i = \frac{\dot{x}^i}{|\dot{x}^i|}$. Consequently, the zeroth component of Eq. (3.4) ($\lambda = 0$) can be written

$$\dot{\varepsilon} + 3\frac{\dot{a}}{a} \varepsilon \dot{x}^i \dot{x}_i = 0. \tag{3.6}$$

The spatial part ($\lambda = i$) of Eq. (3.4) contracted with the vector n_i has the form

$$\ddot{x}^i n_i + 3\frac{\dot{a}}{a} \dot{x}^i n_i \left(1 - \dot{x}^i \dot{x}_i \right) = \left(1 - \dot{x}^i \dot{x}_i \right) k^i_1 n_i + \left(1 - \dot{x}^i \dot{x}_i \right) k^i_2 n_i, \tag{3.7}$$

where $k^i_\alpha = \frac{\partial \xi^i_\alpha}{\partial s_\alpha}$.

Now it is possible to obtain averaged equations [1, 2]. One introduces two macroscopic (averaged) quantities, the energy density and the root-mean-squared (RMS) velocity

$$\frac{E}{V} = \rho = \frac{\sigma_w a^2}{V} \int \varepsilon d^2\sigma, \qquad v^2 = \frac{\int \dot{x}^2 \varepsilon d^2\sigma}{\int \varepsilon d^2\sigma}, \tag{3.8}$$

and can thus average Eqs. (3.6–3.7), obtaining

$$\frac{d\rho}{dt} = -H\rho \left(1 + 3v^2 \right),$$

$$\frac{dv}{dt} = \left(1 - v^2 \right) \left(\frac{K_1 + K_2}{L} - 3Hv \right), \tag{3.9}$$

where t is a physical time, and $H = \dot{a}/a$ is the Hubble parameter, and we made the assumption that curvature radii have the same averaged value and are equal to the correlation length: $R_1 = R_2 = L$. The K_1 and K_2 parameters are curvature/momentum parameters. The component K_1 can be written as $K_1 = u_i' n_i$, suitably averaged over the network, with an analogous definition for K_2.

An evolving wall network loses energy because of possible intersections and the creation of sphere-like objects that eventually collapse. This energy loss mechanism can be added to Eq. (3.9) by analogy to what was originally done by Kibble for cosmic strings [3]. Taking into account this energy loss term, we can rewrite Eq. (3.9) in terms of the correlation length $L = \sigma_w/\rho$, as follows

$$\frac{dL}{dt} = (1 + 3v^2)HL + c_w v,$$
$$\frac{dv}{dt} = (1 - v^2)\left(\frac{k_w}{L} - 3Hv\right), \tag{3.10}$$

where we further defined $k_w = K_1 + K_2$ as the *momentum* parameter.

The momentum parameter can be estimated in an analogous way to what was done for cosmic strings in Ref. [4]. One finds that $k(v)$ can be written similarly as

$$k(v) = k_0 \frac{1 - (qv^2)^\beta}{1 + (qv^2)^\beta}, \tag{3.11}$$

where β, k_0 and q are unknown parameters. The constant k_0 characterizes the maximum value of the momentum parameter: it is positive, but cannot be bigger than 2. The parameter $1/q$ is an averaged maximal velocity for the wall network, which can be shown to be $v_w^2 = 2/3$, as expected, but this requires assumptions that need not be satisfied. In that case the maximal averaged velocity of the network can be smaller (but not larger). As a result we have $0 < 1/q \leq v_w$. Other than these general physical constraints, these parameters must be calibrated numerically.

Energy losses due to scalar radiation were considered in [5]. One finds that the uniformly moving wall does not radiate: only perturbations on the wall surface produce scalar radiation. We have already estimated the level of perturbations through the momentum parameter $k(v)$. The maximal value k_0 corresponds to the minimal RMS velocity and hence to minimal perturbations on the wall surface. Conversely the case when the momentum parameter is zero corresponds to a maximal RMS velocity and a maximally perturbed surface. We thus expect that the amount of radiation is proportional to the surface perturbations, leading to

$$F(v) = c_w v + d[k_0 - k(v)]^r, \tag{3.12}$$

where d and r are constants. In the maximally perturbed (slow expansion) limit $v^2 \to 1/q$ this behaves as

$$F(v) = \frac{c_w}{\sqrt{q}} + dk_0^r, \tag{3.13}$$

and we expect the scalar radiation term to be the dominant one. Conversely in the uniform surface (fast expansion) limit we have

$$F(v) \sim c_w v + d(2k_0)^r q^{\beta r} v^{2\beta r}, \tag{3.14}$$

and in this case we expect the chopping term to be more important, and possibly dominate. Putting together these extensions, the VOS model equations (Eq. 3.10) can finally be rewritten as

$$\frac{dL}{dt} = (1 + 3v^2)HL + c_w v + d[k_0 - k(v)]^r,$$
$$\frac{dv}{dt} = (1 - v^2)\left(\frac{k(v)}{L} - 3Hv\right), \tag{3.15}$$

where $k(v)$ is defined by Eq. (3.11).

3.2   Scaling Solutions

We now discuss all relevant scaling solutions for domain walls [6]. We start by neglecting the effect of the wall density on the background (specifically, on the Friedmann equations). As we shall shortly see this is not a good approximation, since the wall network will generally end up dominating the energy density of the universe. However it is this scenario that is effectively considered, for example, when one performs numerical simulations of domain wall networks.

In this case the attractor solution to the evolution equations (3.15) also corresponds to a linear scaling solution

$$L = \varepsilon t, \qquad v = const. \tag{3.16}$$

Assuming that the scale factor behaves as $a \propto t^\lambda$ the detailed form of the above linear scaling constants is

$$\varepsilon^2 = \frac{k_w(k_w + c_w)}{3\lambda(1 - \lambda)}, \qquad v^2 = \frac{1 - \lambda}{3\lambda}\frac{k_w}{k_w + c_w}. \tag{3.17}$$

As in the case of cosmic strings [7], an energy loss mechanism (that is, a non-zero c_w) may not be needed in order to have linear scaling: by considering the $c_w \to 0$ limit one finds that for $\lambda > 1/4$ a linear scaling solution is always possible. Hence in this case a linear scaling solution may exist in both matter and radiation eras (in the case of cosmic strings this is only guaranteed to be the case in the matter era.) On the other hand, if $\lambda \leq 1/4$ then an energy loss mechanism is necessary to have linear scaling.

Note that the linear scaling solutions are physically very different for cosmic strings and domain walls. In the case of cosmic strings, in the linear scaling phase the string density is a constant fraction of the background density, whereas in the case of domain walls we have $\rho_w/\rho_b \propto t$ so the wall density grows relative to the background density, and will eventually become dominant. This happens at a time $t_\star \sim (G\sigma)^{-1}$. Since the domain wall mass per unit area is related to the energy scale of the phase transition, $\sigma \sim \eta^3$, we can also write out a given epoch as

$$\frac{t_\star}{t_{Pl}} \sim \left(\frac{\eta}{m_{Pl}}\right)^{-3}; \tag{3.18}$$

hence walls that would become dominant around today would have been formed at a phase transition with an energy scale

$$\eta_0 \sim 100\,\text{MeV}; \tag{3.19}$$

notice that this is *two orders of magnitude larger* than the standard Zel'dovich–Kobzarev–Okun bound [8]. It will be seen from the discussion that follows that networks that are much heavier would have become dominant well before having reached the linear scaling regime, whereas networks that are much lighter would not yet have reached the linear regime by today. Hence the range of cosmological scenarios where the linear scaling solution is of interest is quite limited.

3.2.1 The Effect of Friction

At early times, in addition to the damping caused by the Hubble expansion, there is a further damping term coming from friction due to particle scattering off the domain walls. This effect can be adequately described by a frictional force per unit area [9]

$$\mathbf{f} = -\frac{\sigma}{\ell_f}\gamma\mathbf{v}, \tag{3.20}$$

where v is the string velocity, defining a friction length scale

$$\ell_f = \frac{\sigma}{N_w T^4} \propto a^4 \tag{3.21}$$

where T is the temperature of the background and N_w is the number of light particles changing their mass across the wall. Just like in the case of cosmic strings discussed in Chap. 2 we can include this term in the evolution equations, which become

$$\frac{dL}{dt} = HL + \frac{L}{\ell_d}v^2 + c_w v \tag{3.22}$$

$$\frac{dv}{dt} = (1 - v^2)\left(\frac{k_w}{L} - \frac{v}{\ell_d}\right), \tag{3.23}$$

where we have defined a damping length scale

$$\frac{1}{\ell_d} = 3H + \frac{1}{\ell_f}. \tag{3.24}$$

Note that since $\ell_f \propto a^4$, the friction term will be dominant at early times, while the Hubble term will dominate at late times, so the late-time linear scaling solution is unchanged. The timescale when Hubble damping dominates over friction (which is also the timescale for the walls to become relativistic) is again $t_\star$ given in Eq. (3.18). Thus we see that domain wall networks will dominate the energy density of the universe even without ever becoming relativistic or reaching the linear scaling regime.

There will be two possible scaling solutions (which are necessarily transient) during the friction-dominated epoch. As in the case of strings [7], these solutions will exist regardless of whether or not the walls interact with each other (that, is, whether c_w is non-zero or vanishes). The conformal stretching solution is

$$L_s \propto a, \quad v_s \propto \frac{\ell_f}{a}; \tag{3.25}$$

for domain walls this gives $v \propto a^3$, whereas for cosmic strings we would have $v \propto a^2$. We emphasize that although the network is being stretched as the scale factor, and is non-relativistic, the velocities are increasing rather fast, due to the effect of the domain wall curvature. This shows that even in the absence of other mechanisms this regime would only be a transient. The only situation where such a stretching regime could persist would be during an inflationary phase, but in that context the much faster expansion is enough to counter the wall velocities and make them decrease. Indeed, in the case of an exponential expansion the solution is

$$L_{inf} \propto a, \quad v_{inf} \propto a^{-1}. \tag{3.26}$$

Following the conformal stretching regime, or right after the formation of the network if it is formed with high enough density, there is a Kibble regime [3], which in the context of the VOS model can be rigorously derived. The scaling solution has precisely the same form for both types of defects

$$L_k \propto \left(\frac{\ell_f}{H}\right)^{1/2}, \quad v_k \propto \left(\ell_f H\right)^{1/2}, \tag{3.27}$$

although of course the friction lengthscale will not have the same form in the two cases. Notice the differences relative to the stretching regime: here the correlation lengths grow much faster, while the velocities grow relatively more slowly. In the stretching regime the walls are typically quite far apart, so there is very little interaction between them—typically less than one per Hubble volume per Hubble time. In the Kibble regime, on the other hand, the walls are so close together that there is a very large number of interactions—in fact there are more than in the case of the linear scaling regime. This enhanced energy loss makes the correlation length grow quite fast. The wall velocities are still non-relativistic and growing, but because regions of the network with higher velocity than average have a larger interaction probability than slower regions (thus leaving the network) the enhanced energy loss is also responsible for making the velocities grow more slowly than in the stretching case. Still the Kibble scaling is also a transient, which in the absence of other mechanisms will necessarily end when the network becomes relativistic.

Even allowing for friction, linear scaling would be an attractor of the above equations if one neglected the effect of the wall density on the expansion of the universe. However, we have seen that in every scaling regime considered the wall density grows relative to the background, so that a wall density term $\rho_w = \sigma/L$ must be included in the Einstein equations. This changes the situation for it is easy to see that the domain wall network will eventually dominate the energy density of the universe (unless some mechanism like a subsequent phase transition were to make it decay and disappear). Thus we again see that linear scaling is of little practical importance, since it is never reached for any cosmologically realistic network.

Since a domain wall network will eventually dominate the energy density of the universe it is important to study the dynamics of the universe in this regime. The expectation [8] is that the domain wall network will again become frozen in comoving coordinates with

$$L \propto a, \quad a \propto t^2. \tag{3.28}$$

In this case the average distance between the walls also grows as t^2 and rapidly becomes greater than the horizon. This will happen at a time that is again given by t_* above. An inertial observer will see domain walls moving away towards the horizon, and as walls fade away the spacetime around the observer will asymptotically approach Minkowski space. Notice that this solution does not depend on c_w—it is valid whether or not the domain walls interact.

3.2.2 Calibrating the Model

One can easily confirm that the extended VOS model given by Eq. (3.15) possess the same scaling behavior as the original one, given by Eq. (3.10). In the extended model we have in principle 6 undefined parameters that should be determined from numerical simulation data [10–12]. By using bootstrapping techniques one finds that the chopping parameter is negligibly small in comparison with the contribution from scalar radiation and may be neglected as a first approximation (specifically, we find $c_w = 0.00 \pm 0.01$), while the other five parameters have the following values [2]

$$d = 0.28 \pm 0.01, \quad r = 1.30 \pm 0.02, \quad \beta = 1.69 \pm 0.08 \tag{3.29}$$

$$k_0 = 1.73 \pm 0.01, \quad q = 4.27 \pm 0.10. \tag{3.30}$$

This has been shown to provide an excellent agreement with the entire range of numerical simulations with a fixed expansion rate λ.

As an additional test, one can carry out analogous field theory simulations of the radiation-matter transition. In this case the scale factor has the following exact analytic expression

$$\frac{a(\tau)}{a_{eq}} = \left(\frac{\tau}{\tau_*}\right)^2 + 2\left(\frac{\tau}{\tau_*}\right), \tag{3.31}$$

where $\tau_* = \tau_{eq}/(\sqrt{2} - 1)$ and the parameters a_{eq} and τ_{eq} are constants denoting the scale factor and conformal time at the epoch of equal radiation and matter densities. This is an important test of the model, and one finds that the analytic model provides an excellent description of the radiation-matter transition. Fitting the phenomenological parameters to the simulations, it has been found that energy losses due to creation of sphere-like objects are typically subdominant in comparison with scalar radiation, except in the case of fast expansion rates. Overall, the extended analytic model can describe both the fixed expansion rate cases and the transition from the radiation to the matter-dominated era. The latter one is an important test of the model, since the network is not scaling during the transition (while the model parameters were calibrated from fixed expansion rate data in the scaling regime).

3.3 Multi-field Models: Frustrated Expectations

Domain wall networks were suggested as a possible provider of the dark energy inferred from astrophysical observations [13]. This possibility is now excluded, but it provides a useful motivation for studying the conditions under which the network reaches scaling or conversely freezes into what is colloquially called a 'frustrated network' [14].

A domain wall network providing dark energy must be dominating the energy density of the universe around the present day, so its energy density must be of the order of the critical density, $\rho_w = \sigma/L_0 \sim \rho_c \sim 1/Gt_0^2$, which provides us with a unique relation between the energy scale of the defects and the present correlation length, namely $L_0 \sim \eta^3/T_0^3 T_{eq}$. The dark energy should be approximately homogeneous and isotropic on cosmological scales or otherwise that would result in strong (unobserved) signatures on the cosmic microwave background. So the product

$$L_0 H_0 \sim \left(\frac{\eta}{30 \, \text{MeV}}\right)^3 \tag{3.32}$$

must be much smaller than unity. Estimating that we need $L_0 < 1\,\text{Mpc} << H^{-1}$, we find again $\eta < 1\,\text{MeV}$. Now, the averaged equation of state of a domain wall network is given by

$$w_w = \frac{1}{3}(3v_w^2 - 2), \tag{3.33}$$

where now v_w is the averaged RMS velocity of the domain wall network. So in order to accelerate the universe with an equation of state in agreement with observations the wall velocities must necessarily be quite small, implying a friction-dominated network, which in turn implies $k < L_0 H_0 << 10^{-4}$. Hence we see that the curvature of the domain walls must unavoidably be very small. Note that for the case of ordinary cosmic strings [3, 4] k is a parameter depending on the defect velocity, whose value increases and closely approaches unity in the limit of small velocities. So the only possibly realistic candidates are non-standard networks.

More robust constraints on domain walls can be obtained from the cosmic microwave background, such as Planck data. Specifically, one can constrain the allowed contribution of the domain walls to the CMB power spectrum [15]. Domain walls are tightly constrained by their temperature power spectrum shape. High-resolution field theory simulations yield the energy-momentum tensor of a network of domain walls in an expanding universe, covering the radiation, matter and late-time Λ-domination eras. Thus the first precise quantitative constraint on the domain wall surface density, is a bound on the energy scale

$$\eta < 0.93\,\text{MeV}, \tag{3.34}$$

at the 95 % CL for the standard Λ-cosmology. This limits the current fraction of the universe's energy density in domain walls to $\Omega_w < 10^{-7}$.

3.3.1 Two-Field Models

We now discuss models described by two real scalar fields, which is the minimum configuration required in order to form networks with junctions. We will discuss in some detail the class of Bazeia–Brito–Losano models (henceforth referred to as the BBL) [16]. The two-field model has the following Lagrangian

$$\mathcal{L} = \frac{1}{2}\sum_{i=1}^{2}(\partial_\mu\phi_i\partial^\mu\phi_i) + V(\phi_i), \tag{3.35}$$

where the ϕ_i are real scalar fields and the potential is

$$V(\phi_i) = \frac{1}{2} \sum_{i=1}^{2} \left(r - \frac{\phi_i^2}{r} \right)^2 + \frac{\varepsilon}{4} \left(\phi_1^4 + \phi_2^4 - 6\phi_1^2\phi_2^2 + 9 \right). \qquad (3.36)$$

where r and ε are two real parameters. This potential has minima at the vertices of a square in the plane (ϕ_1, ϕ_2), the orientation of which depends on the value of the perturbation parameter ε which varies in the range $-2 < \varepsilon r^2 < 1$. There are a total of six independent topological sectors connecting the different minima. In the range $-1/2 < \varepsilon r^2 < 1$ the minima are

$$\phi_i^2 = \frac{r^2}{1 - \varepsilon r^2}, \quad i = 1, 2, \qquad (3.37)$$

while the range $-2 < \varepsilon r^2 < -1/2$ the minima are

$$\phi_i^2 = \frac{r^2}{1 + \varepsilon r^2/2}, \quad \phi_{j\neq i}^2 = 0. \qquad (3.38)$$

This model allows for Y-type and X-type junctions depending on the value of ε, which also controls the tension of the walls connecting each pair of vacua. There are two classes of walls which we denote *edges* and *diagonals*. In the former the wall joins two neighboring minima in field space, and there are four such walls. In the latter the wall joins two opposite minima in field space, and there are two such walls. In the limit of small ε the ratio of the *diagonal* and *edge* tensions is

$$\frac{\sigma_d}{\sigma_e} = \frac{2 + 3\varepsilon}{1 + 21\varepsilon/8}. \qquad (3.39)$$

Depending on whether the diagonal walls have a tension smaller or larger than twice that of the edge ones, the formation of a Y-type or X-type junction will be favored on energetic grounds. In other words, in the case $\sigma_d < 2\sigma_e$ (which corresponds to $\varepsilon > 0$) we have only Y-type stable junctions, while for $\sigma_d > 2\sigma_e$ (which occurs for $\varepsilon < 0$) only X-type stable junctions will be formed.

There are two values of εr^2 for which $\sigma_d = 2\sigma_e$: $\varepsilon r^2 = 0$ and $\varepsilon r^2 = -1$. Thus for $\varepsilon r^2 > 0$ and $\varepsilon r^2 < -1$ Y-type junctions are favored while for $-1 < \varepsilon r^2 < 0$ X-type junctions are preferred. In passing, we note that the parameter ε determines not only the ratio of the energies in the two sectors, but also influences, among other things, how fast the unstable junctions will decay into the stable ones. These expectations have been confirmed numerically. Note that the particular cases $\varepsilon r^2 = 0$ and $\varepsilon r^2 = -1$ do not represent marginal cases between the formation of the junctions of the type Y and X. Instead, these choices decouple the two fields, and so no junctions are formed anymore.

For an interesting example of a marginal case where both types of junctions are allowed, we can consider the model described by a complex scalar field Φ with Lagrangian

$$\mathcal{L} = \partial_\mu \Phi \partial^\mu \bar{\Phi} - \kappa \left| \Phi^N - 1 \right|^2 ; \tag{3.40}$$

where κ is a real parameter and N is integer. This model has vacua at

$$\Phi = e^{i\frac{n}{N}}, \quad n = 0, 1, \ldots, N - 1. \tag{3.41}$$

We can alternatively define the field's phase as ϕ, and it is then obvious that we will have N minima, evenly spaced around ϕ. The case $N = 2$ produces standard domain walls and $N = 3$ produces Y-type junctions, but the case $N = 4$ is slightly more subtle. The potential (3.40) has supersymmetric properties, and hence the energy of a specific solution depends only on the initial and final vacua. In other words, the possible ways of connecting two opposite vacua (directly or through the intermediate vacuum) will have the same energy. Hence this case is an example of the scenario where $\sigma_d = 2\sigma_e$. There is therefore no local energetic argument preferring one type of junction to the other, and consequently Y-type and X-type junctions will always co-exist.

3.3.2 Three-Field Models

This can easily be extended to the case of models with three scalar fields. We will again consider the analogous BBL model [16], whose Lagrangian is

$$\mathcal{L} = \frac{1}{2} \sum_{i=1}^{3} (\partial_\mu \phi_i \partial^\mu \phi_i) + V(\phi_i), \tag{3.42}$$

where the ϕ_i are real scalar fields and the potential has the form

$$V(\phi_i) = \frac{1}{2} \sum_{i=1}^{3} \left[\left(r - \frac{\phi_i^2}{r} \right)^2 + \varepsilon \left(\phi_i^4 + \frac{9}{2} \right) \right] - 3\varepsilon \left(\phi_1^2 \phi_2^2 + \phi_1^2 \phi_3^2 + \phi_2^2 \phi_3^2 \right) ; \tag{3.43}$$

and again r and ε are two real parameters. As in the two-field case, there are two branches for the minima. In this case for $-2/5 < \varepsilon r^2 < 1/2$ the minima are of the form

$$\phi_i^2 = \frac{r^2}{1 - 2\varepsilon r^2}, \quad i = 1, 2, \tag{3.44}$$

while for $-1 < \varepsilon r^2 < -2/5$ they are of the form

$$\phi_i^2 = \frac{r^2}{1 + \varepsilon r^2}, \quad \phi_{j \neq i}^2 = 0. \tag{3.45}$$

In the former case there are 8 minima, which are placed at the vertices of a cube in the space (ϕ_1, ϕ_2, ϕ_3). In the latter one, there are 6 minima, which are placed at the vertices of an octahedron. The minima of the second case can alternatively be thought of as being located at the centers of the faces of a cube.

In the first case (3.44), there are twenty-eight topological sectors and three kinds of walls, which for obvious reasons we can refer to as *edges*, *external diagonals* and *internal diagonals*. The number of different walls of each type is respectively twelve, twelve and four. In the second case (3.45), there are fifteen topological sectors and two kinds of walls, which we can refer to as *edges* and *axes*. The number of different walls of each type is respectively twelve and three.

Again the choice of the parameter ε determines what type of junctions will be present. For $\varepsilon r^2 > -2/5$, corresponding to the cubic solution of Eq. (3.44), X-type junctions survive only if the junctions involving walls from the diagonal sectors are energetically disfavored. This requires $\sigma_{id} > 3\sigma_e$ (for the internal diagonals) and $\sigma_{ed} > 2\sigma_e$ (for the external ones). These conditions are only verified for $-2/5 < \varepsilon r^2 < 0$, so in this range we do have stable X-type junctions. Conversely, for $0 < \varepsilon r^2 < 1/2$ only the Y-type junctions survive, and these may involve either of the *diagonal* sectors, so we effectively have two types of Y-junctions. For $\varepsilon r^2 < -2/5$, corresponding to the octahedral solution of Eq. (3.45), both Y-type and X-type junctions can survive. Note that in this octahedron branch X-type junctions, as well as any Y-type junctions which involve the *axes* sector (as opposed to only the *edge* sector), correspond to field space configurations where one of the three fields vanishes. The case $\varepsilon = 0$ corresponds to the evolution of the three decoupled fields, with no junctions being formed.

3.3.3   The Ideal Model

The availability of a large number of models (of which the two studied above are but examples), all different but to some extent related, begs the question of which features are fundamental and which are irrelevant details. More to the point, one might ask what is the best possible model for domain walls, at least from the point of view of its potential to produce frustrated networks.

In an ideal model, the probability that two domains in the same vacuum state are close to each other should be as small as possible, since that will minimize the probability of coarse-graining. This can be accomplished in models with a very large number of different vacua (say, having the number of scalar fields involved $N \to \infty$). On the other hand, all the possible domain walls should have equal tensions: if that were not the case we would be adding a different source of instability since the walls with higher tension would tend to collapse.

The ideal model (as far as frustration is concerned) is therefore a model with a very large number of vacua with all the domain walls connecting the various vacua having the same tension. Due to energetic considerations, the stable junctions of such a model must necessarily be of Y-type only. Geometrically speaking, the ideal potential is therefore described by N real scalar fields and has mutually equidistant minima. The number of vacua is $N + 1$ for N real scalar fields, and the energetic cost for a specific transition between any two of them is the same. An explicit realization of such a model is given by the potential [14]

$$
V = \frac{\lambda}{N+1} \sum_{j=1}^{N+1} r_j^2 \left(r_j^2 - r_0^2 \right)^2, \quad r_j^2 = \sum_{i=1}^{N} (\phi_i - p_{i_j})^2, \tag{3.46}
$$

where p_{i_j} are the $N + 1$ coordinates of the vacua of the potential. The p_{i_j} are chosen to be the vertices of an $(N + 1)$-dimensional regular polygon, and the distance between the vacua is given by r_0. The model is therefore the sum of $N + 1$ ϕ^6 potentials. Each of them has one minimum located at the center, and a continuum set of minima at a distance r_0 from the center. Note that N of these vacua are located exactly at the centers of the other potentials. However, numerical simulations show that even in these cases the networks do not frustrate but approach linear scaling [14].

3.4 Biased Walls

We now quantify whether (and, if so, how) the linear scaling solution breaks down in several alternative scenarios, where the standard initial conditions are biased in one of several ways; this analysis must necessarily rely both on analytic calculations and on high-resolution field theory simulations.

A scalar field ϕ with Lagrangian density

$$
\mathcal{L} = \frac{1}{2} (\partial_\mu \phi)(\partial^\mu \phi) - V_0 \left(\frac{\phi^2}{\phi_0^2} - 1 \right)^2, \tag{3.47}
$$

will have domain wall solutions, with the height of the potential barrier and surface tension being respectively

$$
V_0 = \frac{\lambda}{4} \phi_0^4, \quad \sigma \sim \sqrt{\lambda} \phi_0^3, \tag{3.48}
$$

while the wall thickness is

$$\delta \sim \frac{\phi_0}{\sqrt{V_0}} \sim (\sqrt{\lambda}\phi_0)^{-1}. \tag{3.49}$$

By the standard variational methods we obtain the field equation of motion (written in terms of physical time t)

$$\frac{\partial^2 \phi}{\partial t^2} + 3H\frac{\partial \phi}{\partial t} - \nabla^2 \phi = -\frac{\partial V}{\partial \phi}. \tag{3.50}$$

where ∇ is the Laplacian in physical coordinates, $H = a^{-1}(da/dt)$ is the Hubble parameter and a is the scale factor, which we assume to vary as $a \propto t^\lambda$; in particular, in the radiation era $\lambda = 1/2$, while in the matter era $\lambda = 2/3$. This can then be extended in three relevant ways.

An extension which preserves scaling occurs if domain walls are produced during an anisotropic phase in the early universe and are subsequently pushed outside the horizon (and freeze-out in comoving coordinates) due to inflation. In this case they will retain the imprints of this anisotropy, which will only be erased once they re-enter the horizon and become relativistic. Indications of this isotropization (and scaling) were suggested in [17], though robust evidence for it was only presented in [18].

3.4.1 Biased Initial Conditions

The usual choice of initial conditions assumes ϕ to be a random variable uniformly distributed between $-\phi_0$ and $+\phi_0$. Thus the fraction of the simulation box that is in either minimum is initially 50 %, and this fraction is maintained by the subsequent evolution. One can, however, bias the initial conditions by changing the above fractions; a previous inflationary phase could again be responsible for this, by creating Hubble volumes with slightly different occupation fractions.

The phenomenological analysis of [19] suggests that for population fractions close to 50 % (i.e., a weak bias) a good fit is provided by

$$\frac{A}{V} \propto \eta^{-1} \exp(-\eta/\eta_c), \tag{3.51}$$

while for a stronger bias

$$\frac{A}{V} \propto \exp(-\eta/\eta_c), \tag{3.52}$$

is sufficient. In both cases η_c provides a characteristic timescale at which the decay starts. Later on, analytic arguments by Hindmarsh suggested that one should expect [20]

$$\frac{A}{V} \propto \eta^{-1} \exp\left[-\kappa \varepsilon^2 \eta^2\right],$$ (3.53)

The subsequent analysis of [22] claims a qualitative agreement with this formula, though no quantitative measure of it is provided.

Recently, larger simulations were carried out [18], with population fractions in the negative and positive minima

$$f_- = \frac{1}{1+b}, \quad f_+ = \frac{b}{1+b},$$ (3.54)

or equivalently

$$\varepsilon = \frac{1-b}{2(1+b)}.$$ (3.55)

These show that the phenomenological formulas of [19] provide very good fits. The fitted values for η_c and the reduced chi-square of the best fit are respectively

$$\frac{1}{\eta_c} = 0.328 \pm 0.002, \quad \chi_\nu^2 = 1.33$$ (3.56)

for the weak bias case $b = 0.8$ and

$$\frac{1}{\eta_c} = 1.359 \pm 0.002, \quad \chi_\nu^2 = 1.18$$ (3.57)

for the strong bias case $b = 0.6$. If we fix the value of ε (corresponding to the value of b being used) in Eq. 3.53 for the analytic formula of Hindmarsh we find much poorer fits: it is statistically clear that the square dependence on conformal time in this fitting formula is incorrect.

3.4.2 Biased Potential

An asymmetry between the two minima of the potential can also be introduced [21, 22]. In this case the volume pressure from the biasing provides an additional mechanism which will affect the dynamics of these walls. A simple tilted potential is

$$V(\phi) = V_0 \left[\left(\frac{\phi^2}{\phi_0^2} - 1\right)^2 + \mu \frac{\phi}{\phi_0}\right],$$ (3.58)

and the asymmetry parameter (or energy difference between the two vacua) is

$$\delta V = 2\mu V_0.$$ (3.59)

For a network with characteristic curvature radius R the surface pressure (from the tension force) and the volume pressure (from the energy difference between the two minima) are

$$p_T = \frac{\sigma}{R}, \qquad p_V = \delta V. \tag{3.60}$$

Depending on the relative importance of these two mechanisms, the walls may be long-lived (as in the standard case) or disappear almost immediately.

At early times the surface tension tends to dominate (due to the small curvature radii), and as long as this is the case we expect a linear scaling regime as in the standard case. When the domains become large enough they will decay: we typically expect this to happen when

$$R \sim \frac{\sigma}{\delta V}, \tag{3.61}$$

and assuming that $R \sim \eta$ this corresponds to

$$\eta \sim \frac{\phi_0}{\mu\sqrt{V_0}}. \tag{3.62}$$

Once the volume pressure becomes significant the walls are expected to move with an acceleration

$$\frac{\delta V}{\sigma} \sim \lambda^{1/2}\mu\phi_0 \tag{3.63}$$

and rapidly disappear. A sufficiently fast decay may allow these networks to avoid the Zel'dovich bound [8].

Low-resolution simulations were performed in [22], and assuming the fitting function

$$\frac{A}{V} \propto \eta^{-1} \exp\left[-\kappa(\mu\eta)^n\right], \tag{3.64}$$

they suggest that a good fit is provided by an exponent $n = 2 \pm 1$. Note that this $n = 2$ case again corresponds to the fitting formula (3.53). More recent high-resolution simulations [18] confirm that the choice $n = 2$ provides good fits; specifically for $\mu = 0.03$

$$\kappa = (6.34 \pm 0.01) \times 10^{-3}, \quad \chi_\nu^2 = 1.05 \tag{3.65}$$

while for $\mu = 0.1$

$$\kappa = (6.36 \pm 0.01) \times 10^{-3}; \quad \chi_\nu^2 = 1.16 \tag{3.66}$$

note that the value of κ is the same (within the uncertainty, which is listed at the one-sigma level) in both cases.

Thus the above analysis leads us to conclude that the decay rate of networks with biased initial conditions differs from that of networks with a biased potential, and in particular only the latter is well described by Hindmarsh's analytic fitting formula

(3.53). The physical reason for this difference between the two scenarios is related to the assumption of a Gaussian ansatz for the field probability distribution. For the biased potential case this is a good approximation, but it is not so when we have biased initial conditions.

3.5 Physical and Invariant Models

Thus far the characteristic length scale L was an invariant quantity—in other words, a measure of the invariant string energy (and hence length). We now discuss how to express the VOS model in terms of a physical length scale [23]. The invariant and physical energies are related through the standard Lorentz factor, $\gamma = (1 - v^2)^{-1/2}$, as follows

$$E_{inv} = \gamma E_{ph}. \qquad (3.67)$$

Since for a defect with an n-dimensional worldsheet $\rho \propto L^{-(4-n)}$, the characteristic length scales are related via

$$L_{ph} = \gamma^{\frac{1}{4-n}} L_{inv}. \qquad (3.68)$$

This length scale is a measure of the total energy content of the network, or (in the context of the VOS model assumption of a single independent characteristic scale) the typical separation between defects. We may instead define a characteristic defect size

$$S_{inv} = \gamma S_{ph}; \qquad (3.69)$$

this would therefore be a characteristic length (or total length) for the strings, and a characteristic area (or total area) for walls. This can then be equivalently expressed in terms of a characteristic radius $S \propto R^{n-1}$, leading to

$$R_{inv} = \gamma^{\frac{1}{n-1}} R_{ph}. \qquad (3.70)$$

Let us consider the standard VOS model, whose evolution equations are

$$(4-n)\frac{dL_{inv}}{dt} = (4-n)HL_{inv} + v^2\frac{L_{inv}}{\ell_d} + cv, \qquad \frac{dv}{dt} = (1-v^2)\left[\frac{k}{R_{inv}} - \frac{v}{\ell_d}\right]. \qquad (3.71)$$

For clarity, we have now explicitly identified the invariant quantities. As previously discussed, for a universe with $a \propto t^\lambda$ (with $0 < \lambda < 1$), these equations have the attractor scaling solution

$$\left(\frac{L}{t}\right)^2 \equiv \varepsilon^2 = \frac{k(k+c)}{n(4-n)\lambda(1-\lambda)} \qquad (3.72)$$

$$v^2 = \frac{4-n}{n}\frac{1-\lambda}{\lambda}\frac{k}{k+c}. \tag{3.73}$$

Note that in this attractor solution frictional damping due to particle scattering is negligible compared to that due to the Hubble expansion. We can now change variables using

$$\frac{d\gamma}{dt} = v\gamma^3\frac{dv}{dt} \tag{3.74}$$

leading to

$$\frac{d(\gamma v)}{dt} = \frac{k\gamma}{R} - \frac{\gamma v}{\ell_d} \tag{3.75}$$

$$(4-n)\frac{dL_{ph}}{dt} = (4-n)HL_{ph} + kv\frac{L_{ph}}{R_{inv}} + \gamma^{\frac{1}{4-n}}cv. \tag{3.76}$$

Finally, noting that in the canonical model R is an invariant quantity which in a one-scale model context is identified as $R_{inv} \equiv L_{inv}$ and transforming it to the physical one, we finally obtain

$$\frac{d(\gamma v)}{dt} = \frac{k\gamma^{1+\frac{1}{4-n}}}{L_{ph}} - \frac{\gamma v}{\ell_d} \tag{3.77}$$

$$(4-n)\frac{dL_{ph}}{dt} = (4-n)HL_{ph} + v(k+c)\gamma^{\frac{1}{4-n}}. \tag{3.78}$$

Note that the damping length scale does not appear in the evolution equation for the physical length scale, but only in the one for the invariant length scale (as well as in the one for the velocity). If we now look for attractor scaling solutions we get

$$\varepsilon_{ph}^2 = \gamma^{\frac{2}{4-n}}\varepsilon_{inv}^2, \qquad (\gamma v)_{ph}^2 = \gamma^2 v_{inv}^2, \tag{3.79}$$

which is trivially correct and consistent given the various definitions above.

We can also confirm how the model parameters c and k behave as one switches between the physical and invariant approaches. Starting with the energy loss term c, one notes that the probability dP that a defect segment will encounter another segment in a time interval dt should be given approximately by

$$dP = -\frac{d\rho}{\rho} = (4-n)\frac{dL_{ph}}{L_{ph}} \sim c_{ph}\frac{vdt}{L_{ph}} \sim c_{ph}\frac{vdt}{\gamma^{\frac{1}{4-n}}L_{inv}}. \tag{3.80}$$

From this we infer that

$$c_{ph} = \gamma^{\frac{1}{4-n}}c_{inv}. \tag{3.81}$$

For the specific case of strings, if $c_{inv} = c_0\gamma^{-1/2}$ (with c_0 being a constant), then it follows that $c_{ph} = \gamma^{1/2}c_{inv} = c_0 = const$. Our analysis shows that analogous results hold regardless of the defect dimensionality. A similar argument can be made for the curvature parameter, leading to

$$k_{ph} = \gamma^{\frac{1}{4-n}} k_{inv}. \tag{3.82}$$

One important point pertaining to the behavior of this parameter is that $k \to 0$ as $v \to 1$. With these relations between the physical and invariant model parameters, we can finally write

$$(4 - n)\frac{dL_{ph}}{dt} = (4 - n)HL_{ph} + (c_{ph} + k_{ph})v \tag{3.83}$$

$$\frac{dv}{dt} = (1 - v^2)\left[\frac{k_{ph}}{L_{ph}} - \frac{v}{\ell_d}\right], \tag{3.84}$$

or equivalently

$$\frac{d(\gamma v)}{dt} = \frac{\gamma k_{ph}}{L_{ph}} - \frac{\gamma v}{\ell_d}, \tag{3.85}$$

which are the evolution equations for the VOS model based on physical rather than invariant parameters.

3.5.1 Contracting Universes

As an application of the model, we now discuss the evolution of defect networks in contracting universes, clarifying and extending the results of [24, 25]. We temporarily ignore the effects of friction due to particle scattering (in other words, assume $\ell_f \to \infty$). The consequences of relaxing this assumption will be discussed in the following sub-section.

The key physical difference between this case and the standard one is that in a contracting phase the Hubble parameter becomes negative—in other words it becomes an acceleration term (rather than a damping term). As a result the velocity will increase and the network will become ultra-relativistic, with $v \to 1$. This is true even though in this limit we expect $k(v) \to 0$. In this case one easily finds from Eq. (3.85) an asymptotic behavior

$$\gamma v \propto a^{-n}. \tag{3.86}$$

On the other hand, from Eq. (3.71) one finds for the invariant length scale

$$L_{inv} \propto a^{\frac{4}{4-n}}, \tag{3.87}$$

which can be re-expressed in terms of the corresponding physical scale

$$L_{ph} = \gamma^{\frac{1}{4-n}} L_{inv} \propto a. \tag{3.88}$$

This last relation agrees with the intuitive expectation that the defect network is being conformally contracted as the universe collapses. Similarly for the characteristic radius for extended defects we have

$$R_{ph} \propto a, \qquad R_{inv} = \gamma^{\frac{1}{n-1}} R_{ph} \propto a^{-\frac{1}{n-1}}. \tag{3.89}$$

We also note that in all cases the network's energy density behaves as

$$\rho \propto L_{inv}^{-(4-n)} \propto a^{-4}; \tag{3.90}$$

again this is to be expected: an ultra-relativistic network behaves as a radiation fluid. An interesting consequence of this is that, even if the defect network eventually dominates the energy density of the universe, the universe's contraction rate will still be radiation-like. In any case, as the temperature rises as approaches that of the defect-forming phase transition we expect the defects to effectively dissolve into the high-density background.

We can further quantify how this asymptotic scaling regime is approached. This corresponds to studying the behavior of Eqs. (3.83) and (3.85) when the c_{ph} and k_{ph} terms provide a small but not entirely negligible contribution. In the former case, assuming for simplicity that in this regime the scale factor behaves as $a \propto (t_c - t)^\lambda$, where t_c is the Big Crunch time and $0 < \lambda < 1$, we find

$$L_{ph} \propto a \left[1 - \frac{c_{ph}}{(1-\lambda)(4-n)} a^{\frac{1}{\lambda}-1} \right], \tag{3.91}$$

and as expected the correction factor approaches unity as $a \to 0$. Similarly for Eq. (3.85) we find the approximate solution

$$\gamma \propto a^{-n} \exp \left[-\frac{k_{ph}}{1-\lambda} a^{\frac{1}{\lambda}-1} \right] \approx a^{-n} \left[1 - \frac{k_{ph}}{1-\lambda} a^{\frac{1}{\lambda}-1} \right]. \tag{3.92}$$

Note that the form of the correction term is quite similar to that for the length scale equation, the two differences being that the γ correction is independent of the defect dimensionality (there's no dependence on n) and that k_{ph} is expected to be negative in this limit and to approach zero as $v \to 1$ while c_{ph} is expected to be a positive constant. Finally we can put the two together using Eq. (3.68), obtaining

$$L_{inv} \propto a^{\frac{4}{4-n}} \left[1 - \frac{c_{ph} - k_{ph}}{(1-\lambda)(4-n)} a^{\frac{1}{\lambda}-1} \right], \tag{3.93}$$

which matches the physical intuition that c_{ph} should be more important than k_{ph} in determining this correction.

3.5.2 Friction Domination

In the previous sub-section the effects of friction due to particle scattering were neglected. This is a reasonable assumption for heavy defects (say, those formed around the GUT scale), but for very light defects (say, those formed around the electroweak scale) friction will dominate over Hubble damping for a considerable period. In what follows we discuss how the above solutions change in this case. We are comparing the Hubble and friction contributions to the damping length scale

$$\frac{1}{\ell_d} = nH + \frac{1}{\ell_f} = nH + \frac{\Theta}{a^{n+1}},\tag{3.94}$$

where for convenience we wrote the friction length scale in terms of the scale factor by introducing a constant parameter Θ which is related to the parameter θ defined in Eq. 2.43. Both of these parameters count the number of effective degrees of freedom which interact with the defect.

The two terms have generically different dependencies on the scale factor: the Hubble term behaves as $H \propto a^{-1/\lambda}$ (for a scale factor $a \propto t^\lambda$) while the friction term behaves as $\ell_f^{-1} \propto a^{-(1+n)}$. Therefore it follows that for a fast expansion or contraction rate, $\lambda(1 + n) > 1$ the Hubble term decays more slowly (and eventually dominates) in an expanding universe, and conversely it grows more slowly in a contracting universe. In the opposite regime of slow expansion or contraction, corresponding to $\lambda(1 + n) < 1$, it is the friction term that decays more slowly in the expanding case (and grows more slowly in a contracting one). Interestingly, the transition between these fast and slow regimes depends on the dimensionality of the defect: it occurs at $\lambda = 1/2$ (that is, the radiation era) for monopoles, at $\lambda = 1/3$ for cosmic strings, and at $\lambda = 1/4$ for domain walls.

It is interesting to study in full generality the behavior of the defect networks in regimes where the friction term dominates, first by considering all possible expansion rates (that is, whether λ is fast or slow) and second by considering both expanding and contracting universes. The interesting result is that not only do the stretching and Kibble regimes still exist, but they are the only possible ones. Neglecting the Hubble damping term and keeping the friction one, the evolution equations are

$$(4 - n)\frac{dL_{inv}}{dt} = (4 - n)HL_{inv} + \Theta\frac{v^2 L_{inv}}{a^{1+n}} + cv\tag{3.95}$$

$$\frac{dv}{dt} = (1 - v^2)\left[\frac{k}{L_{inv}} - \Theta\frac{v}{a^{1+n}}\right].\tag{3.96}$$

The invariant index in the lengthscale L was kept for completeness, although in this case the physical and invariant lengths are effectively the same: it is easy to show that the only possible attractor solutions of this system have decreasing non-relativistic speeds ($v \to 0$), and therefore the Lorentz factor approaches unity. From this one can then find the two possible solutions. For low density, slow networks the energy loss (chopping term) is negligible and the network is simply conformally stretched (or contracted) as the universe evolves. This is therefore the stretching regime

$$L \propto a, \qquad v \propto \frac{\ell_f}{L} \propto a^n. \tag{3.97}$$

Note that although the defect velocity is small, it is growing if the universe is expanding, and decreasing if the universe is collapsing. This implies that this solution must always be a transient one. If the chopping term cannot be neglected we have the Kibble regime [3]

$$L \propto \sqrt{a^{1+n+\frac{1}{\lambda}}} \propto \sqrt{\frac{\ell_f}{|H|}}, \qquad v \propto \frac{\ell_f}{L} \propto \sqrt{a^{1+n-\frac{1}{\lambda}}} \propto \sqrt{\ell_f |H|}. \tag{3.98}$$

In this case the correlation length is the geometric mean between the damping length and the horizon length. Note that for fast expansion rates $\lambda(1 + n) > 1$ the velocity increases in the standard case of an expanding universe (in which case the solution is a transient one, and is followed by linear scaling) but it would be an attractor for a contracting universe. Conversely, for slow expansion rates $\lambda(1 + n) < 1$ the velocity decreases in an expanding universe (in agreement with the fact that the friction term does dominate asymptotically in this case) whereas it increases in a collapsing universe, meaning this solution would be transient.

References

1. C.J.A.P. Martins, E.P.S. Shellard, Phys. Rev. D **54**, 2535 (1996)
2. C.J.A.P. Martins, I.Y. Rybak, A. Avgoustidis, E.P.S. Shellard, Phys. Rev. D **93**, 043534 (2016)
3. T. W. B. Kibble, Nucl. Phys. **B252**, 227 (1985). (**B261**, 750 (1986))
4. C.J.A.P. Martins, E.P.S. Shellard, Phys. Rev. D **65**, 043514 (2002)
5. T. Vachaspati, A.E. Everett, A. Vilenkin, Phys. Rev. D **30**, 2046 (1984)
6. P.P. Avelino, C.J.A.P. Martins, J.C.R.E. Oliveira, Phys. Rev. D **72**, 083506 (2005)
7. C.J.A.P. Martins, Phys. Rev. D **70**, 107302 (2004)
8. Ya.B. Zel'dovich, I. Kobzarev, L.B. Okun, Soviet Phys. JETP, **40**, 1 (1975)
9. A. Vilenkin, E.P.S. Shellard, *Cosmic Strings and Other Topological Defects* (Cambridge University Press, Cambridge, 1994)
10. W.H. Press, B.S. Ryden, D.N. Spergel, Ap. J. **347**, 590 (1994)
11. A.M.M. Leite, C.J.A.P. Martins, Phys. Rev. D **84**, 103523 (2011)
12. A.M.M. Leite, C.J.A.P. Martins, E.P.S. Shellard, Phys. Lett. B **718**, 740 (2013)
13. M. Bucher, D.N. Spergel, Phys. Rev. D **60**, 043505 (1999)
14. P.P. Avelino, C.J.A.P. Martins, J. Menezes, R. Menezes, J.C.R.E. Oliveira, Phys. Rev. D **78**, 103508 (2008)
15. A. Lazanu, C.J.A.P. Martins, E.P.S. Shellard, Phys. Lett. B **747**, 426 (2015)
16. D. Bazeia, F.A. Brito, L. Losano, Europhys. Lett. **76**, 374 (2006)
17. P.P. Avelino, C.J.A.P. Martins, Phys. Rev. D **62**, 103510 (2000)
18. J.R.C.C.C. Correia, I.S.C.R. Leite, C.J.A.P. Martins, Phys. Rev. D **90**, 023521 (2014)

19. D. Coulson, Z. Lalak, B.A. Ovrut, Phys. Rev. D **53**, 4237 (1996)
20. M. Hindmarsh, Phys. Rev. Lett. **77**, 4495 (1996)
21. G.B. Gelmini, M. Gleiser, E.W. Kolb, Phys. Rev. D **39**, 1558 (1989)
22. S.E. Larsson, S. Sarkar, P.L. White, Phys. Rev. D **55**, 5129 (1997)
23. C.J.A.P. Martins, M.M.P.V.P. Cabral, Phys. Rev. D **93**, 043542 (2016)
24. P.P. Avelino, C.J.A.P. Martins, C. Santos, E.P.S. Shellard, Phys. Rev. Lett. **89**, 271301 (2002)
25. L. Sousa, P.P. Avelino, Phys. Rev. D **83**, 103507 (2011)

Chapter 4
The Rest of the Zoo

Abstract We present an extension of the velocity-dependent one-scale model suitable for describing the evolution of networks of local and global monopoles, including the cases where these are attached to various numbers of strings. We discuss the key dynamical features that need to be accounted for, in particular the fact that the driving force is due to the other monopoles (rather than being due to local curvature as in the case of extended objects) and new forms of energy loss terms due to monopole-antimonopole capture and annihilation. We find that in many cases the networks evolve towards the standard scaling solution but other scaling laws can also exist, depending on the number of strings involved, the universe's expansion rate and the network's energy loss mechanisms. We also study the particularly interesting case of semilocal string networks.

4.1 Monopole Networks

The general physical properties of monopoles have been described in [1–3]. We start by describing the extensions to the VOS model necessary to account for these specific properties, and will also provide a preliminary calibration for the case of global networks [4].

Consider a network of defects with n-dimensional worldsheets ($n = 1$ for monopoles) evolving in $(3 + 1)$ space-time dimensions. Temporarily assume them to have velocity v, to be non-interacting and (for extended objects) planar. Then the momentum per unit comoving defect volume—simply the momentum, for monopoles—goes as

$$p \propto a^{-1} \implies v\gamma \propto a^{-n} \tag{4.1}$$

from which we get by differentiation

$$\frac{dv}{dt} + nH(1 - v^2)v = 0. \tag{4.2}$$

© The Author(s) 2016
C.J.A.P. Martins, *Defect Evolution in Cosmology and Condensed Matter*,
SpringerBriefs in Physics, DOI 10.1007/978-3-319-44553-3_4

On the other hand, under the above hypotheses the average number of defects in a fixed comoving volume should be conserved, which implies

$$\rho \propto \gamma a^{-(4-n)} \tag{4.3}$$

and again, differentiating and using the velocity equation, we get

$$\frac{d\rho}{dt} + H[(4-n) + nv^2]\rho = 0. \tag{4.4}$$

The hypotheses so far are unrealistic, but we can use this as a starting point to build an accurate model. The validity of this process can be checked for the case of cosmic strings and domain walls, where more rigorous derivations have been presented in the previous two chapters.

Let us start by defining a characteristic length scale

$$L^{4-n} = \frac{M}{\rho}, \tag{4.5}$$

where M will have dimensions appropriate for the defect in question (i.e., monopole mass, string mass per unit length, or wall mass per unit area), and can also be written $M \sim \eta^n$, with η being the symmetry breaking scale. Also, we interpret the velocity as being the RMS velocity of the defect network, and allow for energy losses due to interactions, which for extended defects can usually be modeled (on dimensional grounds) by

$$\frac{d\rho}{dt} = -c\frac{v}{L}\rho; \tag{4.6}$$

we will see below that this ρ dependence also applies to global monopoles while for local monopoles the energy loss is proportional to ρ^2. More importantly, defects will be slowed down by friction due to particle scattering, which can be characterized by a length scale

$$\mathbf{f} = -\frac{M}{\ell_f}\gamma\mathbf{v} \tag{4.7}$$

where we are defining

$$\ell_f \equiv \frac{M}{\theta T^{n+1}} \propto a^{n+1} \tag{4.8}$$

and θ is again a parameter counting the number of particle species (or degrees of freedom) interacting with the defect. We can also define an overall damping length which includes both the effect of Hubble damping and that friction due to particle scattering

$$\frac{1}{\ell_d} = nH + \frac{1}{\ell_f}. \tag{4.9}$$

As previously discussed the friction length scale will in most circumstances grow faster than the Hubble length, therefore friction will be dominant at early times, while Hubble damping will dominate at sufficiently late times.

Putting together all of the above effects, we find the following evolution equation for the characteristic length scale L and RMS velocity v

$$(4-n)\frac{dL}{dt} = (4-n)HL + v^2 \frac{L}{\ell_d} + cv \tag{4.10}$$

$$\frac{dv}{dt} = (1-v^2)\left(f - \frac{v}{\ell_d}\right) \tag{4.11}$$

where in the latter we have included the possibility of further driving forces affecting the defect dynamics. Note that f has the units of acceleration: it is the force per unit mass. For extended objects (walls and strings) we already know that this driving force is the local curvature

$$f \sim \frac{k}{L}; \tag{4.12}$$

we are implicitly assuming that our characteristic length scale is the same as the defect curvature radius. For monopoles the situation is more complicated, since there are forces due to other monopoles. The force between a pair of local monopoles has the following form

$$f_{local} \sim \frac{k}{\eta L^2}. \tag{4.13}$$

For global monopoles the force is independent of distance, but note that their mass grows proportionally to the distance; therefore the acceleration is in fact inversely proportional to distance,

$$f_{global} \sim \frac{k}{L}. \tag{4.14}$$

We see that they are in some sense like local strings: a monopole will be effectively heavier when seen on larger scales, and its acceleration should therefore be correspondingly smaller.

The next issue is the fact that there can be many monopoles in a Hubble volume, so the various forces acting on a given one will partially cancel each other. A simple way in which one can try to model this is as a $1/\sqrt{N}$ effect. In other words, the acceleration f becomes $f/\sqrt{N}$, where the number of defects N in a Hubble volume d_H^3 is given by

$$N_g = \left(\frac{d_H}{L}\right)^3. \tag{4.15}$$

This accurately models global monopoles. For the local case, the existence of anti-correlations in the positions of monopoles and anti-monopoles implies that the number of defects is approximately given by [5]

$$N_l \sim \left(\frac{d_H}{L}\right)^2. \tag{4.16}$$

This is to be expected: since the nearest neighbor to a monopole is likely to be an antimonopole (and vice-versa), typically the attractive forces between nearby pairs will be larger than in the uncorrelated case. In other words, the cancellation mechanism is less strong, which is equivalent to saying that the effective number of neighbors is smaller.

Finally, there is the issue of energy losses due to monopole annihilations. The generic form given by Eq. (4.6) is valid for the case of global monopoles (again, these are in some sense like local strings). For a single monopole-antimonopole pair $\dot{\rho} \propto -\rho/R$. This is another way of saying that the timescale for energy losses corresponds (in the fundamental units we are using) to the lengthscale R. On the assumption that the two lengthscales are comparable $R \sim L$ and this matches Eq. (4.6) apart from the allowance for generic velocities. Note that here we are assuming that the separation between monopole-antimonopole pairs is comparable to the network's characteristic lengthscale. Although the two need not be the same, this is a valid assumption in the context of the simple one-scale model we are considering.

In the local case the Coulomb forces between the monopoles and antimonopoles lead to a different energy loss rate. A detailed study by Preskill which implicitly allows for the effect of the anti-correlations [6], leads to the following evolution equation for the monopole number density

$$\frac{dn_M}{dt} + 3Hn_M = -C\eta^{p-2}\frac{n_M^2}{T^p}, \tag{4.17}$$

where C is a dimensionless constant. On physical grounds we should expect that $p \leq 3$. In terms of the correlation length this has the form

$$3\frac{dL}{dt} = 3HL + \frac{A}{L^2 T^p}. \tag{4.18}$$

As for the specific values of p, one physically expects a short, transient high-temperature regime where $p = 2$, and a longer low-temperature regime with $p = 9/10$. This is important because if $p < 1$ annihilations are expected to become unimportant, in which case we expect $n \propto T^3$, which corresponds to $L \propto t^{1/2}$. On the other hand, if $p > 1$ then annihilations are always relevant, and in that case one expects $n \propto T^{p+2}$, which corresponds to $L \propto t^{(p+2)/6}$. (Notice that this analysis is for the radiation dominated epoch.) It is therefore important to see if we can recover these results using the model in the local case.

4.1.1 Evolution of Local Monopoles

We can now look for scaling solutions for the characteristic lengthscale and RMS velocity of the monopoles. Starting with the case of local monopoles, the evolution equations are

$$3\frac{dL}{dt} = 3HL + v^2\frac{L}{\ell_d} + \frac{C\eta^{p-2}}{L^2 T^p} \tag{4.19}$$

$$\frac{dv}{dt} = (1 - v^2)\left(\frac{k}{\eta L^2}\frac{L}{d_H} - \frac{v}{\ell_d}\right). \tag{4.20}$$

We can start by finding solutions in Minkowski space-time, by setting $H = 0$. In this case the asymptotic scaling solution has the form

$$L^3 \propto \frac{\eta^{p-2}}{T^p}t \tag{4.21}$$

and the monopoles will freeze, with the scaling law for the velocity depending on the behavior of the friction length scale. If we have $\ell_f = const$ we find

$$v \propto t^{-4/3}, \tag{4.22}$$

while for the arguably more realistic $\ell_f \propto L$ the freezing happens more slowly,

$$v \propto t^{-1}. \tag{4.23}$$

In the (unrealistic) frictionless limit $\ell_f \to \infty$ the correlation length still has the same scaling, but velocities asymptote to the speed of light, $v \to 1$.

For an expanding universe with $a \propto t^\lambda$, there are two possible scaling laws, which depend on the values of p and on λ. For the case

$$p < 3 - \frac{1}{\lambda} \tag{4.24}$$

we will have

$$L \propto a; \tag{4.25}$$

here energy losses due to annihilation are unimportant and the monopoles are conformally stretched. Note that in the radiation era we do recover $L \propto t^{1/2}$, and in that case the threshold is indeed $p < 1$. This recovers and generalizes the Preskill results. In the opposite case

$$p > 3 - \frac{1}{\lambda} \tag{4.26}$$

annihilations are dynamically important, and the scaling law is

$$L \propto t^{(\lambda p+1)/3};$$

(4.27)

again we recover the expected Preskill result $L \propto t^{(p+2)/6}$ for the radiation era. Note that for $p > 3 - 1/\lambda$ we have $(\lambda p + 1)/3 > \lambda$: as expected, in this regime the evolution is faster than the above (conformal stretching) one. This difference illustrates the effect of the annihilations.

Interestingly, in both regimes the scaling law for the velocities is the same, namely

$$v \propto t^{-\lambda} \propto a^{-1} \propto T.$$

(4.28)

This is a nice and simple result, and it disproves the naive expectation that the monopoles should move with thermal velocities (which would correspond to $v \propto \sqrt{T}$).

It is also worth pointing out that linear scaling ($L \propto t$) will occur for the case $\lambda p = 2$, while for $\lambda p > 2$ the correlation length will grow superluminally. In the latter case annihilations are so efficient that on average there will eventually be less than one monopole per Hubble volume. Since we physically expect $p \leq 3$, then $\lambda p = 2$ corresponds to $\lambda \geq 2/3$. Hence linear scaling can occur in the matter era but not in the radiation era. On the other hand, superluminal scaling requires $\lambda > 2/3$ and so it can't occur in either epoch—this is a simple manifestation of the monopole problem in standard cosmology.

The above solutions hold for decelerating universes (with $0 < \lambda < 1$) but also for power-law inflating universes (that is, with $\lambda \geq 1$). On the other hand, in de Sitter space (with $a \propto e^{Ht}$) we have

$$L \propto a, \quad p \leq 3$$

(4.29)

$$L \propto a^{p/3}, \quad p > 3,$$

(4.30)

although we expect that the latter behavior for p is physically unrealistic. For the velocity we still have

$$v \propto a^{-1}.$$

(4.31)

Thus in an inflating universe the monopoles freeze and are conformally stretched, being pushed outside the horizon. After the end of inflation there will be much less than one monopole per horizon and their velocities will be infinitesimal, so they will keep being conformally stretched until they re-enter the horizon. In order to solve the monopole problem one needs sufficient e-folds of inflation to ensure that the monopoles have not yet re-entered.

4.1.2 Evolution of Global Monopoles

A similar analysis can be done for the global case. We shall see that the different force and energy loss terms will lead to very different scaling laws. This case is also interesting because numerical simulations (albeit low-resolution ones) exist against which we can compare our results. In this case the evolution equations will have the general form

$$3\frac{dL}{dt} = 3HL + v^2\frac{L}{\ell_d} + cv \tag{4.32}$$

$$\frac{dv}{dt} = (1 - v^2)\left[\frac{k}{L}\left(\frac{L}{d_H}\right)^{3/2} - \frac{v}{\ell_d}\right]. \tag{4.33}$$

Again we can start with the Minkowski space-time case. In the unrealistic frictionless limit $\ell_f \to \infty$ we now have asymptotically

$$L = \frac{1}{3}ct, \quad v = 1 \tag{4.34}$$

so global monopoles will become ultra-relativistic. The case of a constant friction length scale is not relevant for global monopoles: due to the linear divergence of their masses, a more realistic situation in Minkowski space time would be that of the friction length scale being proportional to the correlation length itself, $\ell_f \propto L$. In that case we find the following scaling law

$$L \equiv \varepsilon t = \frac{1}{3}v_0(v_0 + c)t, \quad v = k\varepsilon^{3/2} = const. \tag{4.35}$$

This behavior is to be contrasted with the case of local monopoles, whose velocities always approach zero (except in the unrealistic frictionless case). Note that in principle any value of the velocity is a possible solution, including the limit $v = 1$. An interesting question is whether the friction will make the monopole velocities stabilize at some fixed value (and if so, how small this is) or if they will still become arbitrarily close to the speed of light. Incidentally, notice that assuming $\ell_f \propto t^\sigma$ in Minkowski space, requiring a linear scaling $L \propto t$ implies $v = const$ and $\sigma = 1$. In other words, no other non-trivial behavior of the friction length would lead to linear scaling for the network.

Now let us consider a generic expansion law $a \propto t^\lambda$. Here, just as for local strings, the only possible scaling law is linear scaling

$$L = \varepsilon t, \quad v = v_0 = const., \tag{4.36}$$

which is what is found in numerical simulations. Just as in the Minkowski case there are two branches of the solution: the velocities may or may not be ultra-relativistic! Firstly, the ultra-relativistic scaling regime is

$$v_0 = 1, \quad \varepsilon = \frac{c}{3 - 4\lambda}; \tag{4.37}$$

note that this can only hold for $\lambda < 3/4$, but is in principle allowed both in the radiation and in the matter eras. Secondly, the more standard (sub-luminal) scaling regime will is

$$\varepsilon = \frac{cv_0}{3(1 - \lambda) - \lambda v_0^2}, \quad \lambda v_0 = k(1 - \lambda)^{3/2} \varepsilon^{1/2}; \tag{4.38}$$

These relations could be solved explicitly for ε and v_0, but the corresponding expressions would not be too illuminating. However, simplified and physically suggestive solutions can be displayed for both limits of the expansion power λ. In the limit $\lambda \to 0$, we have

$$\varepsilon = \frac{1}{3}cv_0. \tag{4.39}$$

Not surprisingly, this is similar to the Minkowski space-time scaling. On the other hand, in the limit $\lambda \to 1$, we find

$$v_0 = \frac{1}{3}ck^2(1 - \lambda)^2. \tag{4.40}$$

Here the scaling velocity becomes arbitrarily small ($v \to 0$) and $L \propto a \propto t$ so asymptotically this is a conformal stretching regime.

Unlike the ultra-relativistic branch, this non-luminal branch can exist for any λ (that is, for any expansion law), though note that there is a constraint on the scaling value of the velocity

$$v_0^2 < 3(\frac{1}{\lambda} - 1); \tag{4.41}$$

this is trivial for $\lambda < 3/4$ (in such cases any scaling value for the velocity is allowed in principle), but restrictive for faster expansion rates. Note in particular that it agrees with the above finding that $v = 1$ is only allowed for $\lambda < 3/4$.

In passing we also note that the linear scaling solution $L \propto t$, $v = const$ will also hold if we consider a friction length $\ell_f \propto L$ instead of the usual scaling with temperature. Even in that case no other scaling solutions exist. The only change is that in the scaling coefficients (ε and v_0 above), we would need to interpret the parameter λ as having a renormalized value, instead of the value given by the expansion rate. We can also find scaling solutions in inflating universes. Here there is a unique solution, both for power-law inflation and for the de Sitter case, namely

$$L \propto a, \quad v \propto a^{-1}; \tag{4.42}$$

the solution is the same as in the local case, and the number of e-folds of inflation required to keep the monopoles outside the horizon by the present day should be the same in both cases.

As was done for domain walls [7], it would be interesting to carry out high-resolution numerical simulations of global monopoles with a range of different expansion rates in order to check these solutions and provide a good calibration for the model. For the moment we can use the results of Bennett and Rhie [8] and of Yamaguchi [9] for the correlation length, and also the latter's for the velocities, in order to make some simple comparisons. We start by translating these results into our scaling parameter ε in the radiation and matter eras, finding

$$\varepsilon_{r,BR} \sim 1.32, \quad \varepsilon_{m,BR} \sim 1.89 \tag{4.43}$$

$$\varepsilon_{r,Y} \sim 1.32, \quad \varepsilon_{m,Y} \sim 1.59. \tag{4.44}$$

Notice the remarkable agreement in the radiation era; even in the matter era the difference is small considering the relatively low resolution and dynamic range of the simulations. Now, since Yamaguchi's measurement is consistent with $v = 1$, let us assume that this is the case and solve for the energy loss parameter c. We then find

$$c_{r,BR} \sim 1.32, \quad c_{m,BR} \sim 0.63 \tag{4.45}$$

$$c_{r,Y} \sim 1.32, \quad c_{m,Y} \sim 0.53; \tag{4.46}$$

notice that there is a factor of two difference between the values of the parameter in the radiation and matter eras, while we would expect to find similar values in both epochs if the model is broadly correct and the parameter c is a constant (or nearly so). On the other hand, if we assume that we are in the sub-luminal branch (using in both cases Yamaguchi's values for the velocities, since Bennett and Rhie don't provide a measurement),

$$c_{r,BR} \sim 1.32, \quad c_{m,BR} \sim 1.35 \tag{4.47}$$

$$c_{r,Y} \sim 1.32, \quad c_{m,Y} \sim 1.15; \tag{4.48}$$

here we can claim a very good agreement, considering the resolution of the simulations. Also in this branch we can compare the scaling values of the velocities; according to the model we expect the ratio between the matter and radiation era scaling values to be

$$\frac{v_m}{v_r} = \left(\frac{\varepsilon_m}{6\varepsilon_r}\right) \sim 0.5, \tag{4.49}$$

while Yamaguchi finds $v_m/v_r \sim 0.8$; again given the error bars we would argue that the agreement is encouraging.

4.2 Monopoles Attached to One String

It is becoming increasingly clear, particularly in the context of models with extra dimensions such as brane inflation, that hybrid defect networks of monopoles connected by strings are common. Here we extend our analysis to the case of monopoles attached to one string (the so-called hybrid networks). They arise in the following symmetry breaking scheme [10]

$$G \to K \times U(1) \to K. \tag{4.50}$$

The first phase transition will lead to the formation of monopoles, and the second produces strings connecting monopole-antimonopole pairs; the corresponding defect masses will be

$$m \sim \frac{4\pi}{e} \eta_m, \qquad \mu \sim 2\pi \eta_s^2, \tag{4.51}$$

while the characteristic monopole radius and string thickness are

$$\delta_m \sim (e\eta_m)^{-1}, \qquad \delta_s \sim (e\eta_s)^{-1}, \tag{4.52}$$

In the simplest models all the (Abelian) magnetic flux of the monopoles is confined into the strings. In this case the monopoles are often dubbed 'beads'. However, in general—and in most realistic—models stable monopoles have unconfined non-Abelian magnetic charges. As in the case of isolated monopoles, the key difference between the two cases is that unconfined magnetic fluxes lead to Coulomb-type magnetic forces between the monopoles.

Up to the second transition the above formalism for plain monopoles applies, but the hybrid network requires a separate treatment [11]. From the point of view of analytic model-building, the hybrid case presents one crucial difference. In the case of isolated monopoles their evolution can be divided into a pre-capture and a post-capture period: captured monopoles effectively decouple from the rest of the network, and most of the radiative losses occur in the captured phase, where monopole-antimonopole pairs are bound and doomed to annihilation. This means that the network's energy losses could be described as losses to bound pairs and we did not need to model radiative losses explicitly. In the present scenario the monopoles are captured *ab initio* (as soon as the strings form) and therefore we will need to rethink the loss terms—as well as to account for the force the strings exert on the monopoles.

The lifetime of a monopole-antimonopole pair is to a large extent determined by the time it takes to dissipate the energy stored in the string, $\varepsilon_s \sim \mu L$ for a string of length L. This is because the energy in the string is typically **larger** than (or at most comparable to) the energy of the monopole. Monopoles are pulled by the strings with a force $F_s \sim \mu \sim \eta_s^2$, while the frictional force acting on them is $F_{fri} \sim \theta T^2 v$; the corresponding friction lengthscale is $\ell_f \equiv M/(\theta T^2) \sim \eta_m/(\theta T^2)$.

If there are non-confined fluxes, accelerating monopoles can also lose energy by radiating gauge quanta. The rate of energy loss is expected to be given by the classical electromagnetism radiation formula

$$\dot{\varepsilon}_{gauge} \sim -\frac{(ga)^2}{6\pi} \sim -\left(\frac{g\mu}{m}\right)^2 \sim -\left(\frac{\mu}{\eta_m}\right)^2 \qquad (4.53)$$

where g is the magnetic charge. This should be compared with the ratio of gravitational radiation losses, $\dot{\varepsilon}_{grav} \sim -G\mu^2$; the ratio of the two is

$$\frac{\dot{\varepsilon}_{grav}}{\dot{\varepsilon}_{gauge}} \sim \left(\frac{\eta_m}{m_{Pl}}\right)^2, \qquad (4.54)$$

so if there are unconfined fluxes the gauge radiation will be dominant (except in the extreme case where the monopoles form at the Planck scale itself). The characteristic timescales for this process to act on a monopole-antimonopole pair attached to a string of length L can therefore be written

$$\tau_{rad} \sim \frac{L}{Q} \qquad (4.55)$$

where $Q_{gauge} = (\eta_s/\eta_m)^2$ for gauge radiation and $Q_{grav} = (\eta_s/m_{Pl})^2$ for gravitational radiation. These radiation losses should be included in the model's evolution equation for the characteristic lengthscale L of the monopoles.

One can also have hybrid networks of global monopoles connected by global strings [10]. Just as in the case of plain monopoles, the scale-dependent monopole mass and the different behavior of the forces between monopoles (long-range rather than Coulomb-type) change the detailed properties of these defects and warrant a separate treatment. The tension of a global string is $F_s \sim 2\pi \eta_s^2 \ln(L/\delta_s)$ so now there's an additional logarithmic correction, while the long-range force between the monopoles is

$$F_m \sim 4\pi \eta_m^2. \qquad (4.56)$$

If $\eta_m \gg \eta_s$ the monopoles initially evolve as if they were free, with $L \sim t$. Note that this scaling law implies that there are some monopole annihilations. The strings become dynamically important when $F_s \sim F_m$, that is

$$\ln(t\eta_s) \sim \frac{\eta_m^2}{\eta_s^2}, \qquad (4.57)$$

at which point they pull the monopole-antimonopole pairs together, and the network is expected to disappear within a Hubble time. This leads to the expectation that the network will typically annihilate shortly after the strings form.

4.2.1   Local Case

In this case the VOS model evolution equations take the form

$$3\frac{dL}{dt} = 3HL + v^2 L \left(H + \frac{\theta T^2}{\eta_m} \right) + Q_\star \tag{4.58}$$

where $Q_\star$ includes the energy loss terms from gauge radiation (if it exists), gravitational radiation and loop production discussed earlier, possibly with some coefficient of order unity, and

$$\frac{dv}{dt} = (1 - v^2) \left[\frac{k_m}{\eta_m L^2} \frac{L}{d_H} + k_s \frac{\eta_s^2}{\eta_m} - v \left(H + \frac{\theta T^2}{\eta_m} \right) \right]. \tag{4.59}$$

The velocity equation now has two accelerating terms, due to the strings and the inter-monopole Coulomb forces, which we parametrize with coefficients k_s and k_m which are expected to be of order unity. An exception is the Abelian case, where there are no Coulomb forces, so $k_m = 0$ in this case.

It's important to realize that friction will play a crucial role in the radiation era, where it is more important than the Hubble damping term (the opposite is true for the matter era). Indeed in the radiation era we can write

$$\frac{1}{\ell_d} = H + \theta \frac{T^2}{\eta_m} = \frac{1}{t} \left(\frac{1}{2} + \theta \frac{m_{Pl}}{\eta_m} \right) \equiv \frac{\lambda_\star}{t} \tag{4.60}$$

where in the last step we have defined an effective coefficient that is usually much larger than unity (except if there were no particles interacting with the string, $\theta = 0$). In other cosmological epochs (in particular the matter-dominated era) the friction term is negligible, and we have $\lambda_\star = \lambda$ (where we are defining $a \propto t^\lambda$) as usual. It is also illuminating to compare the various terms in the velocity equation. The ratio of the monopole and string accelerating terms is

$$\frac{F_m}{F_s} = \frac{k_m}{k_s} \frac{1}{L d_H \eta_s^2} \sim \left(\frac{\delta_s}{L} \right) \left(\frac{\delta_m}{d_H} \right) \tag{4.61}$$

which is always much less than unity (except if k_s happened to be extremely small). The Coulomb forces are always negligible relative to the string forces. This was expected, since as we pointed out earlier the energy in the strings is typically larger than that in the monopoles. A useful consequence is that we do not need to treat the Abelian and non-Abelian cases separately. Now let us compare the damping and string acceleration terms

$$\frac{F_d}{F_s} = \frac{\lambda_\star}{k_s} \left(\frac{\eta_m}{m_{Pl}} \right) \left(\frac{T}{\eta_s} \right)^2 v \sim \frac{\theta}{k_s} \left(\frac{T}{\eta_s} \right)^2 v \tag{4.62}$$

where in the last step we assumed that we are in the radiation era. Since the evolution of the monopoles before string formation leads to a scaling law $v \propto a^{-1}$ for monopole velocities [4], which are therefore extremely small when the strings form, the above ratio is always less than unity.

This analysis therefore quantitatively confirms the naive expectation that as soon as the strings are formed the string acceleration term will dominate the dynamics and drive the monopole velocity to unity. Recalling that the initial monopole velocity can effectively be taken to be zero, we can write the following approximate solution for the monopole velocity

$$\ln \frac{1+v}{1-v} = 2f_s(t - t_s) \tag{4.63}$$

and we can compare the epoch at which the monopoles become relativistic (t_c) with that of the string-forming phase-transition (t_s), finding

$$\frac{t_c}{t_s} \sim 1 + \frac{\eta_m}{m_{Pl}}, \tag{4.64}$$

so they become relativistic less than a Hubble time after the epoch of string formation; notice that this ratio depends only on the energy scale of the monopoles, and not on that of strings.

We can now proceed to look for solutions for the characteristic monopole length-scale L, assuming for simplicity that $v = 1$ throughout. There are two possibilities. There is a standard linear scaling solution

$$L = \frac{Q_\star}{3(1 - \lambda) - \lambda_\star} t \tag{4.65}$$

in which the energy loss terms effectively dominate the dynamics. In this case the monopole density decays slowly relative to the background density,

$$\frac{\rho_m}{\rho_b} \propto t^{-1}. \tag{4.66}$$

This scaling law can in principle exist for any cosmological epoch provided $\lambda < 3/4$ (for example, in the matter era we have $L = 3Q_\star t$). However, in the radiation epoch we would need the unrealistic $\theta m_{Pl}/\eta_m < 2$, which effectively would mean that friction is absent ($\theta = 0$). The alternative scaling solution, for epochs when friction dominates over Hubble damping (such as the radiation era) and also for any epoch with $\lambda \geq 3/4$ (even without friction) has L growing with a power $\alpha > 1$ given by

$$\alpha = \lambda + \frac{1}{3}\lambda_\star = \frac{1}{3}\left(4\lambda + \theta \frac{m_{Pl}}{\eta_m}\right); \tag{4.67}$$

if there is no friction the scaling law can simply be written

$$L \propto a^{4/3}, \tag{4.68}$$

from which one sees that although the Hubble term is dominant, the scaling is faster than conformal stretching ($L \propto a$) because the velocities are ultra-relativistic. The important point here is that in this regime L grows faster than $L \propto t$, and the number of monopoles per Hubble volume correspondingly decreases.

At the phenomenological level of our one-scale model, this corresponds to the annihilation and disappearance of the monopole network. We can easily estimate the timescale for this annihilation: it will occur when the number of monopoles per Hubble volume drops below unity (or equivalently $L > d_H$). Assuming a lengthscale $L_s = st_s$ at the epoch of string formation (note that the evolution of the monopole network before the stings form is such that s can be much smaller than unity), we find

$$\frac{t_a}{t_s} \sim 1 + \frac{3}{\theta} \left(\frac{2}{s} - 1 \right) \frac{\eta_m}{m_{Pl}}, \tag{4.69}$$

which is comparable to the estimate we obtained using the velocity equation, Eq. (4.64). This is a much faster timescale than the one associated with radiative losses, which can therefore be consistently neglected in this case. There's nothing unphysical about this 'superluminal' behavior, as explained in [12]. We do have a physical constraint that the timescale for the monopole disappearance should not be smaller than the (initial) length of the string segments, $t_a \geq L_s$, but this should always be the case for the above solutions.

This analysis shows that the monopoles must annihilate during the radiation epoch, if one wants to solve the monopole problem by invoking nothing but a subsequent string-forming phase transition. Any monopoles that survived into the matter era would probably be around today. It also shows that monopoles can also be diluted by a sufficiently fast expansion period. Inflation is of course a trivial example of this, but even a slower expansion rate $3/4 < \lambda < 1$ would be sufficient, provided it is long enough to prevent the monopoles from coming back inside the horizon by the present time.

4.2.2 Global Case

In the global case the evolution equations will be

$$3 \frac{dL}{dt} = 3HL + v^2 \frac{L}{\ell_d} + c_\star v \tag{4.70}$$

$$\frac{dv}{dt} = (1 - v^2) \left[\frac{k_m}{L} \left(\frac{L}{d_H} \right)^{3/2} + \frac{k_s}{L} \frac{\eta_s^2}{\eta_m^2} \ln \frac{L}{\delta_s} - \frac{v}{\ell_d} \right], \tag{4.71}$$

and the scale dependence of the monopole mass and string tension imply that this case differs in two ways from the local case. Firstly, the damping term at the string formation epoch (assumed to be in the radiation epoch) can be written

$$\frac{1}{\ell_d} = H + \theta\,\frac{T^2}{\eta_m^2 L} \sim \frac{1}{t}\left[\frac{1}{2} + \theta\left(\frac{\eta_s}{\eta_m}\right)^2\right]. \tag{4.72}$$

The friction term is now subdominant relative to Hubble damping, and of course it decreases faster than it. Friction can therefore be ignored in the analysis. Secondly, the monopole acceleration due to the (global) strings now has an extra logarithmic correction, but more importantly the logarithmically divergent monopole mass again implies that this force is inversely proportional to L (as opposed to being constant in the local case).

Starting again with the velocity equation, the ratio of the string and monopole acceleration terms is now

$$\frac{F_s}{F_m} \sim \frac{k_s}{k_m}\frac{\eta_s^2}{\eta_m^2}\frac{d_H^{3/2}}{L^{3/2}}\ln{(L\eta_s)}, \tag{4.73}$$

and in particular at the epoch of string formation we have

$$\left(\frac{F_s}{F_m}\right)_{t_s} \sim \frac{k_s}{k_m}\frac{\eta_s^2}{\eta_m^2}\ln\frac{m_{Pl}}{\eta_s}, \tag{4.74}$$

so now F_s is initially sub-dominant, except if we happen to have $\eta_s \sim \eta_m$. Moreover, given that $L \propto t$ with a proportionality factor not much smaller than unity, while monopoles evolve freely (before the effect of the strings is important), the ratio will only grow logarithmically, and so the effect of the strings might not be felt for a very long time. Specifically this should happen at an epoch

$$\frac{t_r}{t_s} \sim \frac{\eta_s}{m_{Pl}}exp\left(\frac{\eta_m^2}{\eta_s^2}\right); \tag{4.75}$$

naturally this is exactly the same as Eq. (4.57).

While strings are dynamically unimportant we have $v = const.$ as for free global monopoles, and even when they become important the velocity will only grow logarithmically towards unity. Hence, although the ultimate asymptotic result is the same in both cases ($v \to 1$), the timescale involved is much larger in the global case. Moreover, recall that in the gauge case the monopole velocities before string formation were necessarily non-relativistic and indeed extremely small, and it is the strings that make them reach relativistic speeds. In the global case this is not so: the monopoles will typically have significant velocities while they are free (although their magnitude depends on model parameters that need to be determined numerically), and therefore the impact of the forces due to the strings is vastly smaller in the global case. All this is due to the fact that the force due to the strings is inversely proportional to L instead of being constant. As for the evolution of L, at early times (before strings become important) we have $L \propto t$ just like for free global monopoles. Eventually the strings push the monopole velocities close(r) to unity, and in this limit the L evolution equation looks just like that for the global case with the particular choices $\lambda_\star = \lambda$ and $Q_\star = c$. However, we must be careful about timescales, since here the approach to $v = 1$ is only asymptotic.

Bearing this in mind, for $\lambda < 3/4$ we will still have linear scaling solution

$$L = \frac{c}{3 - 4\lambda}t; \tag{4.76}$$

notice that this is exactly the ultra-relativistic ($v = 1$) monopole scaling solution. Therefore, if the network happened to be evolving in the other (subluminal) linear scaling solution, the only role of the strings would be to gradually switch the evolution to the ultra-relativistic branch. This scaling law can in principle occur both in the radiation and in the matter eras, and it follows that in this case the monopoles will not disappear at all, but will continue to scale indefinitely (with a constant number per Hubble volume). In this case an analysis in terms of the length of each string segment would have to take into account an initial distribution of lengths. Moreover, since the initial velocities need not be ultrarelativistic, the larger segments (which will have smaller coherent velocities) should grow at early times, while the smaller ones will shrink. The decay time will obviously depend on the initial size. We note that such a behavior has been seen in numerical simulations of semilocal strings [13].

The alternative scaling solution, for any epoch with $\lambda \geq 3/4$, is

$$L \propto a^{4/3}, \tag{4.77}$$

and again corresponds to the annihilation and disappearance of the monopole network, which would occur at

$$\frac{t_a}{t_r} \sim \frac{1}{[(1 - \lambda)r]^{3/(4\lambda - 3)}} \tag{4.78}$$

for an initial lengthscale $L_s = rt_r$; given that we now expect r to be not much smaller than unity, this is likely to be very soon after the onset of this scaling regime—very soon after the velocities become ultra-relativistic. Notice that in the local case the monopoles became ultra-relativistic very soon after the strings formed ($t_c \sim t_s$), but in the present case $t_c \gg t_s$ except if the free monopoles were already ultra-relativistic.

4.3 Monopoles Attached to More than One String

We now study defect networks where monopoles are attached to two or more strings. Networks of the first type are commonly called cosmic necklaces, and we refer to networks of monopoles attached to three or more strings as cosmic lattices. The behavior of necklaces and lattices is qualitatively similar, and will be for the most part treated together, through we will point out the small existing differences. But their evolution does differ in several key aspects from both that of individual monopoles and that of monopoles attached to a single string: the main difference is that necklaces and lattices form stable, long-lived networks which usually reach a scaling solution [14].

The defect networks of interest form via the symmetry breaking pattern $G \rightarrow K \times U(1) \rightarrow K \times Z_N$. If G is a semi-simple group, the first phase transition produces monopoles while in the second each monopole becomes attached to N strings. If K is trivial all the (Abelian) magnetic flux of the monopoles is confined into the strings, and there are no unconfined fluxes. However, unconfined non-Abelian magnetic fluxes can exist in the generic case. The previously discussed hybrid case corresponds to $N = 1$, while $N = 2$ corresponds to cosmic necklaces and $N \geq 3$ to cosmic lattices. The corresponding defect masses, monopole radius and string thickness are still as defined in the previous sections: $m \sim (4\pi/e)\eta_m$ and $\mu \sim 2\pi \eta_s^2$, $\delta_m \sim (e\eta_m)^{-1}$ and $\delta_s \sim (e\eta_s)^{-1}$. There are also scenarios where the intermediate phase transition is absent, $G \rightarrow K \times Z_N$, in which case an analogous network still forms but the role of the monopoles is now played by solitons that are usually called 'beads'. In this case the two energy scales are obviously similar, that is $\eta_s \sim \eta_m$.

Again, up to the second transition (if it exists) the models for plain monopoles apply, but once the strings form a separate treatment is needed. If all the strings attached to each monopole have the same tension (which we will assume to be the case) then all the strings pull it with equal forces, and therefore there is no tendency for a monopole to be captured by the nearest antimonopole, unless their separation is of order δ_s. If there are N strings attached to each monopole, its proper acceleration is given by the vector sum of the tension forces exerted by the strings. At a back-of-the-envelope level, each force is of order $f \sim \mu$, and hence one expects that $a \sim \mu/m$. Monopoles should therefore be accelerated to relativistic speeds provided that the characteristic length of string segments, L_s, is such that $\mu L_s \gg m$, that is $L_s/\delta_s \gg \eta_m/\eta_s$.

Aryal et al. [15] first studied the formation and statistical properties of these networks, for $N = 2$ and $N = 3$, showing that for $N \geq 3$ a single network is formed. In all cases they find that the system is dominated by one infinite network comprising more than 90 % of the string length. Some finite networks and closed loops do exist, in numbers rapidly decreasing with their size. Finally, most of the string segments have a length comparable to the typical distance between monopoles (much larger segments being exponentially suppressed). This justifies our assumption of an inter-monopole separation, L_m, comparable to L_s.

In this case the VOS model evolution equations for the monopoles are

$$3\frac{dL}{dt} = (3 + v^2)HL + Q_\star \tag{4.79}$$

$$\frac{dv}{dt} = (1 - v^2)\left(k_s\frac{\eta_s^2}{\eta_m} - Hv\right). \tag{4.80}$$

We have neglected the term due to friction in both equations—it's easy to show this is subdominant at late times. The energy loss term $Q_\star$ in the lengthscale equation is a renormalized quantity, accounting for the various losses present (including string intercommutings, radiation and annihilations). There may be a velocity-dependence of some of these contributions, but as we shall see monopoles will typically have ultra-relativistic velocities $v \sim 1$ and therefore this dependence can be neglected. The velocity equation includes the force due to the strings (with the phenomenological curvature parameter k_s) but we have neglected that due to monopoles, since if it exists (which is only the case for unconfined fluxes) it's always smaller than that due to the strings. Indeed, using the above definitions of mass scales and thicknesses one finds that the ratio of the two forces is

$$\frac{f_m}{f_s} \sim \frac{k_m}{k_s}\left(\frac{\delta_s}{L}\right)^2 \ll 1; \tag{4.81}$$

we expect the k_i to be (dimensionless) coefficients of order unity, though note that they should be different for necklaces and lattices.

From the velocity equation we confirm that the monopole velocities will be driven towards unity, $v \to 1$. As for the monopole length scale, assuming a generic expansion rate $a \propto t^\lambda$, we find different regimes for slow and fast expansion rates

$$L = \frac{Q_\star}{3 - 4\lambda}t, \quad \lambda < 3/4 \tag{4.82}$$

$$L \propto a^{4/3} \propto t^{4\lambda/3}, \quad \lambda \geq 3/4; \tag{4.83}$$

the former explicitly requires a non-zero energy loss rate. We therefore have linear scaling both in the radiation and matter eras. For monopoles $L \propto t$ corresponds to the monopole density decreasing relative to that of the background. However, for fast expansion rates the growth is superluminal, and the network will eventually disappear.

Its easy to establish a link between this analysis and that of Vachaspati and Vilenkin [16]. Since $\rho_m = mn = m/L^3$ and the monopole equation of state is $3p = v_m^2\rho$, we get by substitution

$$3\frac{dL}{dt} = (3 + v^2)HL + \frac{L}{\eta_m}w. \tag{4.84}$$

Now, Vachaspati and Vilenkin are assuming energy losses through gauge radiation; noting that

$$w \sim \frac{(ga)^2}{6\pi} \sim \left(\frac{\mu}{\eta_m}\right)^2 \sim \dot{\varepsilon}_{gauge} \tag{4.85}$$

$$Q_{gauge} \sim L\frac{\dot{\varepsilon}_{gauge}}{\varepsilon_{gauge}} \sim \frac{L}{\eta_m}w \tag{4.86}$$

we see that this evolution equation for L is exactly the same as Eq. (4.79), matching the Q terms (which in our case can phenomenologically account for further energy loss channels). We can also write the analogous equation for strings. In this case $\rho_s = \mu n^{2/3} = \mu/L^2$ and the string equation of state is $3p = (2v_s^2 - 1)\rho$; again we find

$$2\frac{dL}{dt} = 2HL(1 + v_s^2) + w\mu \tag{4.87}$$

$$w\mu \sim \left(\frac{\eta_s}{\eta_m}\right)^2 \sim Q. \tag{4.88}$$

Note the interesting fact that the dimensionless parameter Q determines the energy loss term for both the strings and the monopoles. The above is the usual evolution equation for the cosmic string correlation length [17, 18], if one assumes a constant string velocity—otherwise the energy loss term should depend linearly on velocity. The scaling solution for the monopoles has already been discussed. For the case of the strings the solution is also the expected linear scaling

$$L = \frac{Q}{2 - 2\lambda(1 + v_s^2)}t, \tag{4.89}$$

for constant velocities and provided $\lambda(1 + v_s^2) < 1$. This solution is an attractor, as in the case of normal strings: for very large lengthscales the string velocity would no longer be a constant (the string velocity evolution equation would drive it to smaller values), and a new equilibrium value with a smaller lengthscale would be reached.

We can also consider more generic scaling solutions of the above equations, allowing for the possibility of zero energy losses ($Q = 0$). We will confirm that scaling ($L \propto t$) generically requires $Q \neq 0$. Starting again with the monopoles, for $Q \neq 0$ we have the two branches of the solution discussed above. For $Q = 0$ the solution is always

$$L \propto a^{4/3} \propto t^{4\lambda/3}; \tag{4.90}$$

note that for $\lambda < 3/4$ the lengthscale grows subluminally while for $\lambda > 3/4$ it grows superluminally. In the absence of radiative energy losses, only a fast enough expansion can dilute the network. We can also compare the evolution of the monopole and background densities

$$\frac{\rho_m}{\rho_b} \sim \frac{\eta_m}{m_{Pl}^2} \frac{t^2}{L^3}. \tag{4.91}$$

For the linear scaling solution $L \propto t$ this has the form

$$\frac{\rho_m}{\rho_b} \sim \frac{1}{Q_\star^3} \left(\frac{\eta_m}{m_{Pl}}\right)^2 \left(\frac{T}{m_{Pl}}\right)^2 \propto \frac{1}{t}. \tag{4.92}$$

From this we see that if gauge radiation is present then the energy density of the network is smaller than the background density. However, if the only radiative channel available is gravitational radiation, in which case $Q_\star = Q_{grav} \sim (\eta_s/m_{Pl})^2$, then the network energy density is in fact the dominant one. For the non-scaling branch the density is

$$\frac{\rho_m}{\rho_b} \propto t^{2-4\lambda}, \tag{4.93}$$

and the behavior depends on the cosmological epoch. During the radiation era the density is a constant fraction of that of the background.

4.4   Semilocal Strings

Semilocal strings were introduced as a minimal extension of the Abelian–Higgs model with two complex scalar fields—instead of just one—that make an $SU(2)$ doublet [19, 20]. This leads to $U(1)$ flux-tube solutions even though the vacuum manifold is simply connected. The strings of this extended model have some similarities with ordinary local $U(1)$ strings, but they are not purely topological and will therefore have different properties. Since they are not topological, they need not be closed or infinite and can have ends. These ends are effectively global monopoles with long-range interactions that can make the segments grow or shrink [21].

The symmetry breaking pattern that leads to the formation of strings in this model is $SU(2)_{global} \times U(1)_{local} \to U(1)_{global}$ so this model can be thought of as a particular limit of the Glashow–Weinberg–Salam electroweak model in which the SU(2) symmetry is global: the Weinberg angle is $\cos \theta_W = 0$ and there are no SU(2) gauge fields. The vacuum manifold is the three sphere, so one would not expect strings to form if the dynamics is dominated by the potential energy. On the other hand, the magnetic field is massive and magnetic flux is conserved, which would suggest the existence of magnetic flux tubes when the magnetic mass is large. This is the regime in which strings form and are stable. The stability of the strings is not trivial, and it will depend on the value of the parameter

$$\beta = \frac{m_{scalar}^2}{m_{gauge}^2}; \tag{4.94}$$

for $\beta < 1$ the string is stable, for $\beta > 1$ it is unstable, and for $\beta = 1$ it is neutrally stable. Only low values of β will be of interest for the purpose of numerical studies, because otherwise the string network is either unstable or disappears very fast.

After a cosmological phase transition in such a model, it is expected that segments of semilocal strings will form. The network evolution depends on the interplay between string dynamics and monopole dynamics. When a string segment ends, it must end in a cloud of gradient energy. Those string ends behave like global monopoles providing an interaction between strings that is independent of distance. Therefore, depending on the interplay between string dynamics and monopole dynamics, the segments can contract and eventually disappear, or they can grow to join a nearby segment and form a very long string, and also the two ends of a segment can join to form a closed loop.

At least to a first approximation, we can envisage these networks as being made of local strings attached to global monopoles, and, as such, the above analytic modeling techniques should be applicable. This being said, it is also clear that these networks possess additional dynamical properties, beyond those of standard hybrid networks. Specifically, the evolution of the string network will depend both on the string tension and on the dynamics of the gradient energy: the latter may be thought of as providing a long-range interaction between the strings. (Note that the force between global monopoles is independent of distance.)

4.4.1 The VOS Model for Semilocal Strings

Our analysis focuses on the behavior of the network as a whole, starting from the premise that it can be treated as a network of local strings attached to global monopoles. Our model for the evolution of these networks is based on explicitly modeling the dynamics and interactions of the monopoles [22, 23]. This is justified since, as has been shown in previous work [13], it is indeed the monopoles that control the evolution of the network.

Recall that the idea is to obtain an evolution equation for the monopole density (neglecting interactions) and then re-express it in terms of a characteristic lengthscale, L (which in this case should be thought of as the average inter-monopole distance). The effects of monopole forces and friction are then included in this equation (as well as in the relevant velocity equation) by adding extra phenomenological terms. The evolution equation for the characteristic monopole lengthscale then has the form

$$3\frac{dL}{dt} = 3HL + v^2\frac{L}{\ell_d} + c_\star v, \qquad (4.95)$$

where $c_\star$ is a free parameter (to be calibrated by simulations) quantifying energy loss, and where as usual we have defined a damping lengthscale,

$$\frac{1}{l_d} = H + \frac{1}{l_f}.$$

(4.96)

The evolution equation for the RMS velocity v of the monopoles is

$$\frac{dv}{dt} = (1 - v^2)\left[\frac{k_m}{L}\left(\frac{L}{d_H}\right)^{3/2} + \frac{k_s}{L}\frac{\eta_s^2}{\eta_m^2} - \frac{v}{\ell_d}\right],$$

(4.97)

where the first term in square brackets is the force due to the monopoles and the second describes the force due to the strings. Note that the fact that the string and monopole symmetry breaking scales appear in Eq. (4.97) is a consequence of the fact that these equations of motion are obtained by modeling semilocal strings as local strings attached to global monopoles, and appropriately adapting the equations of motion for both. Physically one knows that it is the monopoles that dominate the semilocal string dynamics, and this can be modeled by assuming that $\eta_s \ll \eta_m$. Similarly, the horizon size d_H enters in the monopole force term due to a number counting argument: this force depends on the number of monopoles (and antimonopoles) inside the horizon [4].

In the semilocal case the ratio of the forces due to strings and monopoles is

$$\frac{f_s}{f_m} = \frac{k_s}{k_m}\left(\frac{\eta_s}{\eta_m}\right)^2\left(\frac{d_H}{L}\right)^{3/2}$$

(4.98)

and since $\eta_s \ll \eta_m$ the string force is always subdominant. This is in agreement with theoretical expectations and numerical simulations. Note that this is a distinguishing characteristic of these networks: as we saw in previous sections, for local strings attached to local monopoles the force due to the strings always dominates the dynamics, while for global strings attached to global monopoles the string force is subdominant at string formation but becomes dominant later in the network's evolution. One interesting consequence of the fact that the monopoles always dominate the dynamics is that the only attractor solution of these evolution equations in an expanding universe (with $a \propto t^\lambda$) is linear scaling. Indeed, one finds that the only consistent asymptotic solution is

$$L = \gamma t, \quad v = v_0,$$

(4.99)

as in the case of plain global monopoles.

There are two possible branches of the scaling solution. First, there is an ultra-relativistic one with

$$\gamma = \frac{c_\star}{3 - 4\lambda}, \quad v_0 = 1,$$

(4.100)

which only exists for slow expansion rates ($\lambda < 3/4$) but is in principle allowed both on the radiation and matter eras. Second, a normal solution exists for any expansion rate, with scaling parameters

$$\gamma = \frac{c_\star v_0}{3 - \lambda(3 + v_0^2)} \tag{4.101}$$

$$\lambda v_0 = k_m (1 - \lambda)^{3/2} \gamma^{1/2}, \tag{4.102}$$

and a constraint on the velocities

$$v_0^2 < 3(\frac{1}{\lambda} - 1). \tag{4.103}$$

This constraint is trivial for $\lambda < 3/4$ (that is, $v_0 \to 1$ is allowed), but restrictive for faster expansion rates. On the other hand, velocities will generically be significant: having $v_0 \to 0$ requires $\lambda \to 1$.

For comparison we also consider the case of Minkowski space (corresponding to $\lambda = 0$ and $H = 0$) but with a friction lengthscale proportional to the correlation length (say, for simplicity, $\ell_f \sim L$). This should be an adequate description of some of the numerical simulations of semilocal strings done so far [13]. In this case, linear scaling is still the attractor solution but the scaling parameters now obey

$$3\gamma = v_0^2 + c v_0, \quad v_0 = k_m \gamma^{3/2}. \tag{4.104}$$

In the opposite limit of fast expansion rate ($\lambda \geq 1$, or in other words inflation) the linear scaling solution of Eq. (4.99) no longer exists. In this case the network is conformally stretched and gradually frozen, and the characteristic lengthscale and velocity evolve as

$$L \propto a, \quad v \propto \frac{1}{HL}. \tag{4.105}$$

These conformal stretching solutions are ubiquitous in the defects literature. Work on modeling semilocal segment evolution has also been done [22], but an accurate description will need better simulations.

4.4.2 Comparing to Simulations

One should proceed with caution if trying to extract quantitative information from these scaling properties. Nevertheless, it is encouraging that the overall behavior of recent high-resolution simulations is in agreement with our understanding of the relevant underlying physical mechanisms [23]. Specifically, we note that

- For a given cosmology (in numerical terms, damping term), the string correlation length grows with β while the monopole separation gets smaller. This is to be expected since for lower β we expect the system to behave more like an Abelian–Higgs network, which has longer strings and fewer segments. Analogous results have recently been found for cosmic strings [24].
- For a given β, the string correlation length is higher for higher damping terms, and the monopole separation is smaller. This is also to be expected since a lower damping term means that monopole velocities will be higher. Segments can therefore move faster to either grow and meet with other segments or collapse, giving a longer typical string length and smaller number of monopoles. One naturally expects that the additional length lost by segment collapse is more than compensated by that gained by the extra growth. Note that increasing the string correlation length L_s corresponds to decreasing the string density, and therefore the total length in string.

We should also point out that the scaling properties we have obtained for the string segments and monopoles are somewhat less sensitive to the value of β than one might have expected. It is possible that this is a feature of the PRS algorithm [25], which is used for all field theory simulations of this type; this has also been discussed in [24].

As pointed out in [23], a full direct calibration of the parameters of the VOS model for the evolution of the overall network cannot be done until we can numerically determine the velocities of the monopoles and segments—a task which is being pursued at the time of writing. Still, we can provide a preliminary comparison with the model, and specifically with the scaling solution described by Eqs. (4.101–4.102). We will neglect the β dependence, which as we saw is numerically found to be quite small when allowing for statistical and systematic uncertainties.

With these assumptions our free parameters are the analytic model parameters, $c_\star$ and k_m, as well as the monopole scaling velocities in the radiation and matter eras, which we will denote v_{rad} and v_{mat}. Using our numerically determined values of the monopole separations one finds

$$v_{\mathrm{rad}} \sim 0.48 k_m \tag{4.106}$$

$$v_{\mathrm{mat}} \sim 0.20 k_m; \tag{4.107}$$

we have deliberately not included error bars in these numbers since it is not possible to quantify possible systematic uncertainties in the monopole separations. These values are consistent with the results of earlier simulations [22], which for a faster expansion rate ($\lambda = 3/4$) found

$$v_{\mathrm{fast}} \sim 0.12 k_m. \tag{4.108}$$

As expected, faster expansion rates lead to smaller velocities. On the assumption that the analytic model is correct, we therefore infer that the ratio of the scaling monopole velocities in the matter and radiation eras should be

$$\frac{v_{\text{mat}}}{v_{\text{rad}}} \sim 0.4. \qquad (4.109)$$

If one assumes a curvature parameter k_m of order unity as in the case of Goto–Nambu strings, our estimated velocities are comparable to (though possibly somewhat lower than) the ones typically encountered in other field theory defect simulations of domain walls and cosmic strings [7, 26]. Thus, even though this comparison is somewhat simplistic, the results are at least encouraging. A full comparison (and thus a proper calibration of the analytic model) requires the numerical implementation of a reliable method to measure defect velocities in our simulations, which is not yet available.

References

1. S. Coleman, The magnetic monopole fifty years later, in A Zichichi et al. ed. *The Unity of Fundamental Interactions* (Plenum Press, New York, 1983)
2. J. Preskill, Ann. Rev. Nucl. Part. Sci. **34**, 461 (1984)
3. A. Vilenkin, E.P.S. Shellard, *Cosmic Strings and other Topological Defects* (Cambridge University Press, Cambridge, 1994)
4. C.J.A.P. Martins, A. Achucarro, Phys. Rev. D **78**, 083541 (2008)
5. M.B. Einhorn, D.L. Stein, D. Toussaint, Phys. Rev. D **21**, 3295 (1980)
6. J. Preskill, Phys. Rev. Lett. **43**, 1365 (1979)
7. C.J.A.P. Martins, I.Y. Rybak, A. Avgoustidis, E.P.S. Shellard, Phys. Rev. D **93**, 043534 (2016)
8. D.P. Bennett, S.H. Rhie, Phys. Rev. Lett. **65**, 1709 (1990)
9. M. Yamaguchi, Phys. Rev. D **65**, 063518 (2002)
10. A. Vilenkin, Nucl. Phys. B **196**, 240 (1982)
11. C.J.A.P. Martins, Phys. Rev. D **80**, 083527 (2009)
12. A.E. Everett, T. Vachaspati, A. Vilenkin, Phys. Rev. D **31**, 1925 (1985)
13. A. Achucarro, P. Salmi, J. Urrestilla, Phys. Rev. D **75**, 121703 (2007)
14. C.J.A.P. Martins, Phys. Rev. D **82**, 067301 (2010)
15. M. Aryal, A.E. Everett, A. Vilenkin, T. Vachaspati, Phys. Rev. D **34**, 434 (1986)
16. T. Vachaspati, A. Vilenkin, Phys. Rev. D **35**, 1131 (1987)
17. C.J.A.P. Martins, E.P.S. Shellard, Phys. Rev. D **53**, 575 (1996)
18. C.J.A.P. Martins, E.P.S. Shellard, Phys. Rev. D **54**, 2535 (1996)
19. M. Hindmarsh, Phys. Rev. Lett. **68**, 1263 (1992)
20. T. Vachaspati, A. Achucarro, Phys. Rev. D **44**, 3067 (1991)
21. A. Achucarro, T. Vachaspati, Phys. Rept. **327**, 347 (2000)
22. A.S. Nunes, A. Avgoustidis, C.J.A.P. Martins, J. Urrestilla, Phys. Rev. D **84**, 063504 (2011)
23. A. Achcarro, A. Avgoustidis, A.M.M. Leite, A. Lopez-Eiguren, C.J.A.P. Martins, A.S. Nunes, J. Urrestilla, Phys. Rev. D **89**, 063503 (2014)
24. T. Hiramatsu, Y. Sendouda, K. Takahashi, D. Yamauchi, C.M. Yoo, Phys. Rev. D **88**, 085021 (2013)
25. W.H. Press, B.S. Ryden, D.N. Spergel, Ap. J. **347**, 590 (1994)
26. J.N. Moore, E.P.S. Shellard, C.J.A.P. Martins, Phys. Rev. D **65**, 023503 (2002)

Chapter 5
Model Extensions

Abstract We extend the VOS model of string evolution to the case of models with additional degrees of freedom on the string worldsheet, studying both the case of superconducting strings and that of wiggly strings. In the former case we obtain the microscopic string equations of motion in the Witten–Carter–Peter superconducting model, finding that whether the standard scale-invariant evolution of the network is preserved or destroyed due to the presence of the charge will depend on the amount of damping and energy losses experienced by the network. This suggests, among other things, that results derived in Minkowski space (field theory) simulations may not extend to the case of an expanding universe. We also model the evolution of realistic wiggly cosmic strings. Here we discuss model solutions in the extreme limit where the wiggles make up a high fraction of the total energy of the string network (the tensionless limit) and also provide a brief discussion of the opposite (linear) limit where wiggles are a small fraction of the total energy. We also use these results to make extrapolations for the case of cosmic superstrings.

5.1 Superconducting Strings

Cosmic strings can have non-trivial internal structure, generally carrying additional degrees of freedom on the string worldsheet. This is the generic situation in models with extra dimensions, where there is a proliferation of scalar fields that can couple to (and condense on) the strings. Cosmic strings with additional worldsheet degrees of freedom (scalar charges, currents, fermionic zero-modes) have been described in field theory and supergravity [1], but the cosmological evolution of such string networks remains comparatively unexplored.

It is therefore desirable to understand how these additional degrees of freedom can be described macroscopically, and how their presence affects the behavior and cosmological consequences of the corresponding string networks. We start with cosmic strings with a conserved charge living on the string worldsheet. We extend the VOS model to describe this case analytically, and study the effect of the charge on the evolution of the string network.

We consider the Witten–Carter–Peter (chiral) model [2, 3], which implicitly makes use of the fact that in two dimensions a conserved current can be written

© The Author(s) 2016
C.J.A.P. Martins, *Defect Evolution in Cosmology and Condensed Matter*,
SpringerBriefs in Physics, DOI 10.1007/978-3-319-44553-3_5

as the derivative of a scalar field. The dynamics is described by the action

$$S_W = \int \sqrt{-\gamma} \left[-\mu_0 + \frac{1}{2}\gamma^{ab}\phi_{,a}\phi_{,b} - qA_\mu x^\mu_{,a}\frac{\tilde{\varepsilon}^{ab}}{\sqrt{-\gamma}}\phi_{,b} \right] d^2\sigma - \frac{1}{16\pi}\int d^4x\sqrt{-g}F_{\mu\nu}F^{\mu\nu},$$

(5.1)

where the four terms are respectively the usual Nambu–Goto term (μ_0 being the string tension and γ the pullback of the background metric g on the worldsheet), the inertia of the charge carriers described by the scalar ϕ, the current coupling to the electromagnetic potential A_μ, and the kinetic term for the electromagnetic field. Worldsheet indices are denoted by $a, b \in \{0, 1\}$ and we will take σ^0 to be the timelike coordinate, while $\sigma^1 \equiv \sigma$ will be spacelike; $\tilde{\varepsilon}^{ab}$ is the alternating tensor 2D dimensions. This action applies to both the bosonic and the fermionic case.

We are interested in the chiral limit of this model, that is (with dot/prime denoting differentiation with respect to the timelike/spacelike worldsheet coordinate σ^0/σ):

$$\phi'^2 = \varepsilon^2\dot{\phi}^2,$$

(5.2)

where ε is the scalar

$$\varepsilon \equiv \frac{-x'^2}{\sqrt{-\gamma}},$$

(5.3)

giving the string energy per unit coordinate length. In an FRW background

$$ds^2 = a^2(d\tau^2 - dx^2)$$

(5.4)

and choosing the standard gauge $\sigma^0 = \tau$, $\dot{\mathbf{x}} \cdot \mathbf{x}' = 0$ in which the scalar ε becomes:

$$\varepsilon = \left(\frac{\mathbf{x}'^2}{1 - \dot{\mathbf{x}}^2}\right)^{1/2},$$

(5.5)

and introducing the simplifying function Φ defined as

$$\Phi(\phi) = \frac{\dot{\phi}^2}{\mu_0 a^2(1 - \dot{\mathbf{x}}^2)},$$

(5.6)

the microscopic equations of motion take the form

$$[\varepsilon(1 + \Phi)]^\cdot + \frac{\varepsilon}{\ell_d}\dot{\mathbf{x}}^2 = \Phi' - 2\frac{\dot{a}}{a}\varepsilon\Phi,$$

(5.7)

$$\varepsilon(1 + \Phi)\ddot{\mathbf{x}} + \frac{\varepsilon}{\ell_d}(1 - \dot{\mathbf{x}}^2)\dot{\mathbf{x}} = \left[(1 - \Phi)\frac{\mathbf{x}'}{\varepsilon}\right]' + \left(\dot{\Phi} + 2\frac{\dot{a}}{a}\Phi\right)\mathbf{x}' + 2\Phi\dot{\mathbf{x}}',$$

(5.8)

where the damping length is defined in the usual way.

The curvature vector $d^2\mathbf{x}/ds^2$, with $ds = \sqrt{\mathbf{x'}^2}d\sigma$, satisfies:

$$\frac{d^2\mathbf{x}}{ds^2} = \frac{\mathbf{x''}}{\mathbf{x'}^2} - \frac{(\mathbf{x'} \cdot \mathbf{x''})\mathbf{x'}}{\mathbf{x'}^4}. \tag{5.9}$$

Thus, a curvature radius, R, can be defined locally via

$$\frac{\dot{\mathbf{x}}}{\varepsilon(1 - \dot{\mathbf{x}}^2)} \cdot \left(\frac{\mathbf{x'}}{\varepsilon}\right)' = -\frac{\mathbf{x'} \cdot \dot{\mathbf{x}}'}{\mathbf{x'}^2} = \frac{\mathbf{x''} \cdot \dot{\mathbf{x}}}{\mathbf{x'}^2} = \frac{a}{R}(\dot{\mathbf{x}} \cdot \mathbf{u}), \tag{5.10}$$

where we have introduced a unit vector $\mathbf{u}$ in the direction of the curvature vector. The worldsheet charge and current densities are given by

$$\rho_w = q\varepsilon\dot{\phi}, \qquad j_w = q\frac{\phi'}{\varepsilon}, \tag{5.11}$$

while the total energy of a piece of string is given by

$$E = \mu_0 a \int (1 + \Phi)\,\varepsilon d\sigma = E_s + E_\Phi. \tag{5.12}$$

We can immediately interpret this as being split in an obvious way into a string component and a charge component. Defining a macroscopic charge as the average of Φ over the string worldsheet, we then have:

$$Q = \langle\Phi\rangle \equiv \frac{\int \Phi\varepsilon d\sigma}{\int \varepsilon d\sigma} = \frac{E_\Phi}{E_s}. \tag{5.13}$$

This interpretation will be relevant below.

Introducing the network string density ρ_s such that $E_s \propto \rho_s a^3$ and defining the correlation length ξ by

$$\rho_s = \frac{\mu_0}{\xi^2}, \tag{5.14}$$

the evolution equations have the form

$$\frac{\dot{E}_s}{E_s} = \frac{\dot{\rho}_s}{\rho_s} + 3\frac{\dot{a}}{a} = -2\frac{\dot{\xi}}{\xi} + 3\frac{\dot{a}}{a} = \frac{\dot{a}}{a} + \left\langle\frac{\dot{\varepsilon}}{\varepsilon}\right\rangle, \tag{5.15}$$

$$v^2 = \langle\dot{\mathbf{x}}^2\rangle, \qquad v\dot{v} = \langle\dot{\mathbf{x}} \cdot \ddot{\mathbf{x}}\rangle \tag{5.16}$$

$$\dot{Q} = \langle\dot{\Phi}\rangle, \tag{5.17}$$

while the lengthscale L is defined by $\rho = \mu_0/L^2$. An additional difficulty which is absent in the case of Nambu–Goto strings is the appearance of a term proportional to $(\mathbf{x}' \cdot \mathbf{x}'')$. This factor is not expected to be zero even though $(\mathbf{x}' \cdot \mathbf{u})$ is, see Eq. (5.9). Finally, note that in the case of the VOS model for plain Nambu–Goto strings one assumes that the network has a single characteristic length scale, so that $R = L = \xi$. This is no longer true in the charged case, but we will still assume that $R = \xi$, while L is now only a measure of the total energy in the network.

In our chiral case of a conserved microscopic charge we have

$$\rho_w = q\varepsilon\dot{\phi} = q\phi' = const. \tag{5.18}$$

and therefore

$$\Phi = \frac{\varphi_0^2}{a^2 \mathbf{x}'^2}, \tag{5.19}$$

with φ_0 being a constant. By simple differentiation one finds that Φ evolves as

$$\dot{\Phi} + 2\frac{\dot{a}}{a}\Phi = 2\Phi \frac{\dot{\mathbf{x}} \cdot \mathbf{x}''}{\mathbf{x}'^2} \tag{5.20}$$

$$\Phi' + 2\Phi \frac{\mathbf{x}' \cdot \mathbf{x}''}{\mathbf{x}'^2} = 0. \tag{5.21}$$

The conserved charge assumption simplifies some of the above equations. In particular, one easily obtains

$$\ddot{\mathbf{x}} \cdot \mathbf{x}' = \frac{\mathbf{x}' \cdot \mathbf{x}''}{\varepsilon^2}, \tag{5.22}$$

$$\varepsilon' = 0. \tag{5.23}$$

An alternative way to see this is to use Eq. (5.8) together with Eqs. (5.20)–(5.21). This leads to

$$\frac{2\Phi}{1 + \Phi}\frac{\varepsilon'}{\varepsilon} = 0, \tag{5.24}$$

thus if $\Phi \neq 0$ we must have $\varepsilon' = 0$.

5.1.1   Averaged Equations

We can now proceed and look at the averaged evolution equations in our conserved microscopic charge case [4–6]. The total energy of the string is given by Eq. (5.12) and therefore we can define two characteristic lengths for the string: the usual correlation length ξ associated with the *string* energy

$$E_s = \mu a \int \varepsilon d\sigma \propto \rho_s a^3, \tag{5.25}$$

through Eq. (5.14), and the lengthscale L associated with the *total* energy $E \propto \rho a^3$, see Eq. (5.12). Taking the time derivative of the previous equation, one easily finds the evolution equation for ξ

$$2\frac{\dot{\xi}}{\xi} = 2H\left(1 + \frac{v^2}{1+Q}\right) + \frac{2Q}{1+Q}\frac{kv}{R} - \left\langle\frac{\Phi'}{\varepsilon(1+\Phi)}\right\rangle,\tag{5.26}$$

where k is the usual momentum parameter. Correspondingly, the evolution equation for L is

$$2\frac{\dot{L}}{L} = 2H\left(1 + \frac{v^2+Q}{1+Q}\right) - \left\langle\frac{\Phi'}{\varepsilon(1+\Phi)}\right\rangle.\tag{5.27}$$

As defined above, the macroscopic charge Q is the averaged microscopic charge $\langle\Phi\rangle$, which is also the ratio between the string energy E_s and the 'charge' energy E_Φ. Differentiating (5.13), we find that its evolution equation is

$$\frac{\dot{Q}}{Q} = 2\left(\frac{kv}{R} - H\right)\tag{5.28}$$

Finally, the string velocity is defined by

$$v^2 = \langle\dot{\mathbf{x}}^2\rangle = \frac{\int \dot{\mathbf{x}}^2 \varepsilon d\sigma}{\int \varepsilon d\sigma},\tag{5.29}$$

and taking time derivatives on both sides, we arrive at the evolution equation

$$\dot{v} = \frac{1-v^2}{1+Q}\left[\frac{k}{R}(1-Q) - 2Hv + \frac{1+Q}{v}\left\langle\frac{\Phi'}{\varepsilon(1+\Phi)}\right\rangle\right].\tag{5.30}$$

Equations (5.26)–(5.28) are not independent: there is a consistency relation

$$E_s = \frac{E}{1+Q} \longrightarrow \xi = L\sqrt{1+Q}\tag{5.31}$$

Therefore, the equations are related by

$$2\frac{\dot{L}}{L} = 2\frac{\dot{\xi}}{\xi} - \frac{\dot{Q}}{1+Q},\tag{5.32}$$

which is verified, so the three equations are consistent.

In order to proceed we now need to deal with the $(\mathbf{x}' \cdot \mathbf{x}'')$ term coming from Φ'. Referring to Eq. (5.21), dimensional analysis suggests an ansatz of the form

$$\left\langle \frac{\Phi'}{\varepsilon(1+\Phi)} \right\rangle = -s\frac{v}{R}\frac{2Q}{1+Q} \qquad (5.33)$$

where s is (at least, to a first approximation) a constant. Using (5.33) and noting our earlier identifications $R = \xi = L\sqrt{1+Q}$, our evolution equations become:

$$2\frac{\dot{\xi}}{\xi} = 2H\left(1 + \frac{v^2}{1+Q}\right) + \frac{2Q}{1+Q}\frac{(k+s)v}{\xi} \qquad (5.34)$$

$$2\frac{\dot{L}}{L} = 2H\left(1 + \frac{v^2+Q}{1+Q}\right) + \frac{2Q}{(1+Q)^{3/2}}\frac{sv}{L} \qquad (5.35)$$

$$\frac{\dot{Q}}{Q} = 2\left(\frac{kv}{\xi} - H\right) \qquad (5.36)$$

$$\dot{v} = \frac{1-v^2}{1+Q}\left[\frac{k}{\xi}(1 - Q(1+2s/k)) - 2Hv\right]. \qquad (5.37)$$

We will assume a flat universe with generic expansion rate $a \propto t^\lambda$, and look for scaling solutions of the form

$$\xi = \xi_0 t^\alpha, \quad v = v_0 t^\beta, \qquad Q = Q_0 t^\gamma. \qquad (5.38)$$

Note that causality implies $\alpha \leq 1$ and the finite speed of light implies $\beta \leq 0$. Furthermore, an analysis of loop solutions shows that $v \to 1$ is not a physically allowed solution for these networks.

5.1.2 No Charge Losses

We assume that there are no macroscopic charge losses, and we will separately consider the cases with and without energy losses due to loop production. Whether or not we have loop production, the evolution equation for the macroscopic charge Q is given by Eq. (5.36), and we can start by studying this. There is a trivial but unphysical solution if $k = 0$, with $\xi \propto L \propto a$ and $v \propto Q \propto a^{-2}$, which can therefore be ignored. In the realistic case $k \neq 0$ there can in principle be two kinds of solutions:

- Decaying charge solutions, with $\gamma = -2\lambda$ and $\beta < \alpha - 1$; for these solutions not only does the charge decay (as $Q \propto a^{-2}$) but velocity will necessarily decay as well.
- standard solutions with $\beta = \alpha - 1$ and $\gamma = -2\lambda + 2kv_0/\xi_0$; here we have used the term 'standard' referring to the fact that linear scaling solution (with $\alpha = 1$ and $\beta = 0$) is of this form, although a priori there is no guarantee that this solution will exist with a constant (non-zero) charge. Also note that in this branch of solutions we may at least in principle have growing, constant, or decaying Q.

We can now study the entire system of equations in the cases with and without energy losses [6]. In the latter case we obtain the following three scaling relations:

- For slow expansion rates, $\lambda < 2/3$,

$$\alpha = \frac{3}{2}\lambda < 1, \quad \xi_0 = \frac{kv_0}{\lambda} \tag{5.39}$$

$$\beta = \alpha - 1 < 0 \tag{5.40}$$

$$\gamma = 0, \quad Q_0 = \left(1 + \frac{2s}{k}\right)^{-1} \tag{5.41}$$

$$\frac{\rho_s}{\rho_{crit}} \propto \frac{\rho}{\rho_{crit}} \propto t^{2-3\lambda}; \tag{5.42}$$

here we have a constant charge, which gradually slows the strings (making $v \to 0$), although the evolution is still faster than conformal stretching ($\alpha = \lambda$). As a consequence, both the energy density in the strings and the total energy density in the network grow relative to that of the cosmological background.

- For $\lambda = 2/3$, corresponding to the matter-dominated era,

$$\alpha = 1, \quad \xi_0 = \frac{3}{2}kv_0 \tag{5.43}$$

$$\beta = 0, \quad v_0^2 = \frac{1}{2}\left[1 - Q_0\left(1 + \frac{2s}{k}\right)\right] \tag{5.44}$$

$$\gamma = 0 \tag{5.45}$$

$$\frac{\rho_s}{\rho_{crit}} = \frac{1}{1 + Q_0}\frac{\rho}{\rho_{crit}} = \frac{16\pi}{3k^2}\left[1 - Q_0\left(1 + \frac{2s}{k}\right)\right]^{-1}G\mu; \tag{5.46}$$

in this case the macroscopic charge is still a constant, but the additional dilution caused by the faster expansion rate is enough to ensure that the energy density of the network is a constant fraction of the background one. Hence we have a generalized linear scaling solution, with ξ (and L) growing as fast as allowed by causality, in which the RMS velocity and the macroscopic charge are constant (and larger charges leading to smaller velocities).

- For fast expansion rates, $\lambda > 2/3$,

$$\alpha = 1, \quad \xi_0^2 = \frac{k^2}{4\lambda(1-\lambda)} \tag{5.47}$$

$$\beta = 0, \quad v_0^2 = \frac{1-\lambda}{\lambda} \tag{5.48}$$

$$\gamma = 4 - 6\lambda < 0 \tag{5.49}$$

$$\frac{\rho_s}{\rho_{crit}} = \frac{\rho}{\rho_{crit}} = \frac{32\pi}{3k^2}\frac{1-\lambda}{\lambda}G\mu; \tag{5.50}$$

here the additional Hubble damping makes the macroscopic charge decay, and we therefore end up with the solution for ordinary Nambu–Goto strings.

If we now assume that there are energy losses through loop production, this can be described by a new term in the evolution equation for $\dot{\xi}$ (and, consequently, one for $\dot{L}$), which has the usual form

$$2\frac{\dot{\xi}}{\xi} = \ldots + \frac{cv}{\xi}. \tag{5.51}$$

This introduces the parameter c quantifying the efficiency of producing loops. Using Eq. (5.32) and assuming there is no charge loss, we also find

$$2\frac{\dot{L}}{L} = \ldots + \frac{cv}{L}\frac{1}{\sqrt{1+Q}}. \tag{5.52}$$

The analysis can now be repeated, and one finds generalized solutions including an additional dependency on c:

- For slow expansion rates, $\lambda < 2/(3 + c/k)$,

$$\alpha = \frac{1}{2}\left(3 + \frac{c}{k}\right)\lambda < 1, \quad \xi_0 = \frac{kv_0}{\lambda} \tag{5.53}$$

$$\beta = \alpha - 1 < 0 \tag{5.54}$$

$$\gamma = 0, \quad Q_0 = \left(1 + \frac{2s}{k}\right)^{-1} \tag{5.55}$$

$$\frac{\rho_s}{\rho_{crit}} \propto \frac{\rho}{\rho_{crit}} \propto t^{2-(3+c/k)\lambda}; \tag{5.56}$$

loop production is an additional energy loss mechanism, and therefore the correlation length now grows faster than in the $c = 0$ case. Similarly, string velocities decrease more slowly, and the network's density grows more slowly relative to the background one.

- For an intermediate expansion rate, $\lambda = 2/(3 + c/k)$,

$$\alpha = 1, \quad \xi_0 = \frac{1}{2} v_0 \, (3k + c) \tag{5.57}$$

$$\beta = 0, \quad v_0^2 = \frac{1}{2} \left[1 - Q_0 \left(1 + \frac{2s}{k} \right) \right] \tag{5.58}$$

$$\gamma = 0 \tag{5.59}$$

$$\frac{\rho_s}{\rho_{crit}} = \frac{1}{1 + Q_0} \frac{\rho}{\rho_{crit}} = \frac{16\pi}{3k^2} \left[1 - Q_0 \left(1 + \frac{2s}{k} \right) \right]^{-1} G\mu ; \tag{5.60}$$

now the expansion rate for which this solution occurs decreases, with the smaller expansion rate being compensated, for the purposes of the dilution of the network's energy density, by the process of loop production. For $c = 0$ this solution exists for the matter era, and $c = k$ will make it occur in the radiation era. Interestingly, the ratio of the string and background energies is exactly the same as before—it does not depend on the value of c.

- For fast expansion rates, $\lambda > 2/(3 + c/k)$,

$$\alpha = 1, \quad \xi_0^2 = \frac{k(k + c)}{4\lambda(1 - \lambda)} \tag{5.61}$$

$$\beta = 0, \quad v_0^2 = \frac{1 - \lambda}{\lambda} \frac{k}{k + c} \tag{5.62}$$

$$\gamma = \frac{4}{k + c} \left[k - \frac{\lambda}{2} (3k + c) \right] < 0 \tag{5.63}$$

$$\frac{\rho_s}{\rho_{crit}} = \frac{\rho}{\rho_{crit}} = \frac{32\pi}{3k(k + c)} \frac{1 - \lambda}{\lambda} G\mu ; \tag{5.64}$$

this is exactly the VOS linear scaling solution for Nambu–Goto strings, with an added prediction of a particular decay law for the charge (which depends on the cosmological expansion rate).

5.1.3 Macroscopic Charge Losses

We now relax the assumption of no charge losses. This section is somewhat more phenomenological since it will rely on simplifying assumptions for how the charge may be lost, but our goal is simply to develop an intuitive picture for the possible role of charge loss mechanisms on the evolution of the network. We will again assume an energy loss term of the form

$$\frac{\dot{E}_s}{E_s} = -c\frac{v}{\xi}, \quad 2\frac{\dot{\xi}}{\xi} = \ldots + c\frac{v}{\xi} \tag{5.65}$$

but this time let us say that L gets a different (in general) loss term

$$\frac{\dot{E}}{E} = -fc\frac{v}{L}, \quad 2\frac{\dot{L}}{L} = \ldots + fc\frac{v}{L}, \tag{5.66}$$

where f is an arbitrary function, possibly of the velocity, charge and correlation length. We are therefore assuming that any such charge losses are related to the network's intercommuting and loop production mechanisms. In other words, while loops of ordinary string networks decay by emitting gravitational waves (and possibly also by particle production), in our case the emission of charged particles provides an additional decay channel that enhances the total energy losses, and this possible enhancement is phenomenologically described by the parameter f.

In the previous section, we assumed there was no charge loss, implicitly using

$$f(v, Q, \xi) = \frac{1}{\sqrt{1 + Q}}. \tag{5.67}$$

Instead, we will now leave f free and obtain the evolution equation for Q from the above assumptions together with Eq. (5.32). The result is

$$\dot{Q} = 2Q\left(\frac{kv}{\xi} - H\right) + \frac{cv}{\xi}(1 + Q)\left(1 - f\sqrt{1 + Q}\right), \tag{5.68}$$

or an analogous equation in terms of L. With (5.67) we trivially recover the results of the previous sub-section. We can now look for scaling relations as before, checking how the results depend on the choice of f. Note that in the standard VOS model case (without a charge), we would not expect charge to be created. Therefore, $\dot{Q} = 0$ when $Q = 0$, implying

$$2\frac{cv}{\xi}(1 - f) = 0. \tag{5.69}$$

It then follows that, if the function f is constant, it has to be equal to unity. Otherwise it must be dependent on the charge (and possibly other quantities), and become unity when the charge drops to zero,

$$f = f(Q, v, \xi) \longrightarrow f(0, v, \xi) = 1 \,, \tag{5.70}$$

as was the case for the function (5.67) we used in the case without charge loss.

Let us therefore consider the $f = 1$ case. The scaling laws now are

- For slow expansion rates, $\lambda < (2k - Wc)/[3k + (1 - W)c]$,

$$\alpha = \frac{3}{2} \left[\frac{1 + (1 - W)c/3k}{1 - Wc/2k} \right] \lambda < 1, \quad \xi_0 = \frac{v_0}{\lambda}(k - Wc/2) \tag{5.71}$$

$$\beta = \alpha - 1 < 0 \tag{5.72}$$

$$\gamma = 0, \quad Q_0 = \left(1 + \frac{2s}{k} \right)^{-1} \tag{5.73}$$

$$\frac{\rho_s}{\rho_{crit}} \propto \frac{\rho}{\rho_{crit}} \propto t^{2(1-\alpha)} \tag{5.74}$$

where for simplicity we have kept α in the last expression and introduced

$$W = (1 + Q_0^{-1}) \left(\sqrt{1 + Q_0} - 1 \right) , \tag{5.75}$$

which is always positive and behaves as $W = 0.5(1 + Q_0)$ in the limit $Q_0 \to 0$ and as $W \propto \sqrt{Q_0}$ for $Q_0 \to \infty$. The scaling exponent for ξ now has an explicit dependence on the charge. Increasing the scaling value of the charge pushes the value of the maximal expansion rate where this regime holds to larger values, whereas the scaling exponent α decreases. Similarly, string velocities decrease faster, and the network's density grows faster relative to the background one. In principle, as one makes W progressively larger, the scaling exponent α becomes closer to λ, which corresponds to the conformal stretching case. However, in practice we do not expect this to occur, since it is clear from its definition that W should be a small parameter in realistic (cosmological) defect networks.

- For an intermediate expansion rate, $\lambda = (2k - Wc)/[3k + (1 - W)c]$,

$$\alpha = 1, \quad \xi_0 = \frac{1}{2}v_0 [3k + (1 - W)c] \tag{5.76}$$

$$\beta = 0, \quad v_0^2 = \frac{1}{2 - Wc/k} \left[1 - Q_0 \left(1 + \frac{2s}{k} \right) \right] \tag{5.77}$$

$$\gamma = 0 \tag{5.78}$$

$$\frac{\rho_s}{\rho_{crit}} = \frac{1}{1 + Q_0} \frac{\rho}{\rho_{crit}} = \frac{16\pi}{3k^2} \left[1 - Q_0 \left(1 + \frac{2s}{k} \right) \right]^{-1} G\mu \,. \tag{5.79}$$

As in the previous solution there is an explicit dependence on the amount of charge, and the expansion rate for which this solution occurs increases with the charge: a larger charge makes scaling harder, requiring more energy losses (from the damping due to the Hubble expansion, or from losses due to loop production) to counteract it. In fact the expansion rate for which this solution exists would approach $\lambda = 1$ as the scaling value of the charge Q_0 becomes arbitrarily large, although as we already pointed out we do not expect this to occur in practice. The ratio of the string and background energies is still exactly the same as before—it does not depend on the value of c or W.

- For fast expansion rates, $\lambda > (2k - W_0 c)/[3k + (1 - W_0)c]$,

$$\alpha = 1, \quad \xi_0^2 = \frac{k(k+c)}{4\lambda(1-\lambda)} \tag{5.80}$$

$$\beta = 0, \quad v_0^2 = \frac{1-\lambda}{\lambda} \frac{k}{k+c} \tag{5.81}$$

$$\gamma = \frac{4}{k+c}\left[k - \frac{W_0 c}{2} - \frac{\lambda}{2}[3k + (1 - W_0)c]\right] < 0 \tag{5.82}$$

$$\frac{\rho_s}{\rho_{crit}} = \frac{\rho}{\rho_{crit}} = \frac{32\pi}{3k(k+c)} \frac{1-\lambda}{\lambda} G\mu; \tag{5.83}$$

which is again the VOS linear scaling solution for Nambu–Goto strings, now with a faster decay law for the charge (which is obvious since we have explicit charge losses). For this solution we used the notation W_0 to indicate the value of W in the limit $Q_0 \to 0$, that is $W = 1/2$; the reason for this choice will become clear below.

It is easy to check that if we set $c = 0$ and/or $W = 0$ we recover the results of the previous section. However, note that the transition between the second and third solutions will now depend on the amount of charge loss. The expansion rate of the second solution coincides with the minimum expansion rate of the third solution for

$$W = W_0 = \frac{1}{2}, \tag{5.84}$$

which corresponds to the $Q_0 = 0$ limit.

Finally, it is interesting to discuss what happens in the more general case where $f \neq 1$. In the absence of compelling arguments suggesting a particular form for f (other than the above case without charge losses) we will consider a linearized form

$$f(Q) = 1 + wQ, \tag{5.85}$$

for w real with $|w| < 1$. The rationale for this is that in realistic networks in cosmological contexts the charges are likely to correspond to a small fraction of the overall energy density of the network. Repeating the analysis we find that the above solutions still hold, provided that one extends the definition of the parameter W to

$$W = (1 + Q_0^{-1}) \left[(1 + wQ_0)\sqrt{1 + Q_0} - 1 \right]. \tag{5.86}$$

In the limit $Q_0 \to 0$ this now behaves as

$$W = \left(\frac{1}{2} + w \right)(1 + Q_0); \tag{5.87}$$

we trivially recover the behavior in the $w = 0$ (constant) case, but we also see that in this limit W will vanish if $w = -1/2$. This is interesting because that choice of w corresponds to the linearized version of $f = 1/\sqrt{1 + Q}$, which as we argued is the case of no charge losses. This shows that the above analysis is self-consistent.

5.2 Wiggly Strings

It is well known that cosmologically realistic string networks are not quite of Goto–Nambu type. Numerical simulations of cosmic strings in expanding universes have established beyond doubt the existence of a significant amount of short-wavelength propagation modes (commonly called *wiggles*) on the strings, on scales that can be several orders of magnitude smaller than the correlation length. This small-scale structure can be optimally described through its fractal properties [7]. On large scales we expect strings to be Brownian (with a fractal dimension $d_{large} \sim 2$), while on small enough scales strings are smooth and locally straight (having $d_{small} \sim 1$). Between these two scales, one finds an intermediate fractal region that extends over several orders of magnitude. This fractal region evolves in time, spreading out between the initial correlation length and the horizon size in such a way that any given physical scale is always loosing power.

We stress that it is still not clear under which conditions small-scale structure continues building up indefinitely or eventually reaches a scaling solution like the large-scale properties of the network, although some progress has been made by a number of authors [8, 9]. Due to the very limited number of degrees of freedom available, the Goto–Nambu model cannot account for this phenomenology. More general string models [10, 11] are extremely useful for this purpose, and the much larger amount of algebra required is generously compensated by the resulting physical phenomenology. Here we summarize the mathematical formalism necessary for a generalized VOS model that explicitly accounts for the build-up of small-scale structures on the strings [4, 12].

The motion of a cosmic string is in general obtainable from a variational principle applied to the action

$$S = - \int \mathscr{L} \sqrt{-\gamma}\, d^2\sigma\,;$$

(5.88)

where the worldsheet metric is given by $\gamma_{ab} = g_{\mu\nu}x^\mu_{,a}x^\nu_{,b}$. Quite generically, string models can be described by a Lagrangian density $\mathscr{L}$ depending only on the space-time metric $g_{\mu\nu}$, background fields such as a Maxwell-type gauge potential A_μ or a Kalb–Ramond gauge field $B_{\mu\nu}$ (but not their gradients) and any relevant internal fields, contained in a function Λ (see below), that is

$$\mathscr{L} = \Lambda + J^\mu A_\mu + \frac{1}{2}W^{\mu\nu}B_{\mu\nu} + \dots\,.$$

(5.89)

The Maxwell and Kalb–Ramond fields are ideal for describing superconducting and global strings respectively, but for wiggly strings they can be set to zero—the effect of small-scale structures can be encoded in the function Λ. The simplest Goto–Nambu string model is obviously

$$\mathscr{L}_{GN} = -\mu_0\,.$$

(5.90)

Models having a variable Lagrangian density are usually called elastic string models [10, 11]. The reason for this is that the energy density in the locally preferred string rest frame, which will henceforth be denoted by U, and the local string tension, denoted T, are constant for a Goto–Nambu string $U = T = \mu_0$ but they are variable in general. In particular, one should expect that the string tension in an elastic model will be reduced with respect to the Goto–Nambu case due to the mechanical effect of the current.

Since elastic string models necessarily possess conserved currents, it is convenient to define a 'stream function' ϕ on the world-sheet that will be constant along the current's flow lines. The part of the Lagrangian density $\mathscr{L}$ containing the internal fields can be defined as a function of the magnitude of the gradient of this stream function, $\Lambda = \Lambda(\chi)$, such that $\chi = \gamma^{ab}\phi_{,a}\phi_{,b}$; notice that in more general cases with non-zero external fields these would be covariant derivatives. In our case, we will require a single scalar field, and the associated current can be pictured as a mass current. This means that *we should think of wiggly strings as carrying a mass current*, which will renormalize the bare mass per unit length μ_0. Indeed, the model with Lagrangian density

$$\mathscr{L} = -\mu_0\sqrt{1-\chi}\,,$$

(5.91)

has the equation of state

$$UT = \mu_0^2\,.$$

(5.92)

and it has been shown that this equation of state arises in an *exact* way [13] in a macroscopic (in the sense of *smoothed*) model of a wiggly string, that is a Goto–Nambu string containing a spectrum of small oscillations that one cannot (or does not want to) describe in microscopic detail. Consistently with this physical picture, we will make the simplifying assumption that the potential χ depends only on the world-sheet time. We are therefore interpreting it as an *effective* or renormalized quantity, defined on a scale that is smaller than the horizon (which is the scale beyond which the network is Brownian) but still large enough (just) for the small-scale dependence on the space-like world-sheet coordinate to be negligible. In this phenomenological sense it can be pictured as a *mesoscopic* quantity.

The free string equations of motion can now be obtained in the usual (variational) way. We retain the line element and gauge choice of previous chapters, and the coordinate energy per unit length along the string is still given by

$$\varepsilon^2 = \frac{\mathbf{x}'^2}{1 - \dot{\mathbf{x}}^2} . \tag{5.93}$$

The only difference (apart from the additional amount of algebra) is that there is now a further equation for the scalar field ϕ. Indeed, rather than working with this directly it turns out to be convenient to define the dimensionless parameter w by

$$\Lambda = -\mu_0 w ; \tag{5.94}$$

and then the local string tension and energy density are simply given by

$$\frac{T}{\mu_0} = w , \qquad \frac{U}{\mu_0} = \frac{1}{w} , \qquad \frac{T}{U} = w^2 . \tag{5.95}$$

Incidentally, notice that the equation of state for these networks has the form

$$3\frac{p}{\rho} = \left(1 + \frac{T}{U}\right) v^2 - \frac{T}{U} . \tag{5.96}$$

Hence wiggly strings still behave as radiation ($p/\rho \sim 1/3$) in the ultra-relativistic limit. On the other hand, in the non-relativistic limit one has

$$\left(\frac{p}{\rho}\right)_{nr} = -\frac{1}{3}\frac{T}{U} \geq -\frac{1}{3} \tag{5.97}$$

while in the tensionless limit ($T/U \to 0$)

$$\left(\frac{p}{\rho}\right)_{nt} = \frac{1}{3}v^2 ; \tag{5.98}$$

tensionless non-relativistic wiggly strings behave as matter ($p/\rho \sim 0$). It may be of interest to assess the possible role of such strings in the context of the dark matter problem.

Thus one finds the following microscopic string equations of motion

$$\ddot{\mathbf{x}} + \dot{\mathbf{x}}(1 - \dot{\mathbf{x}}^2)\frac{\dot{a}}{a}(1 + w^2) = \frac{w^2}{\varepsilon}\left(\frac{\mathbf{x}'}{\varepsilon}\right)', \tag{5.99}$$

$$\left(\frac{\varepsilon}{w}\right)' + \left(\frac{\varepsilon}{w}\right)\frac{\dot{a}}{a}\left[2w^2\dot{\mathbf{x}}^2 + (1 + \dot{\mathbf{x}}^2)(1 - w^2)\right] = 0, \tag{5.100}$$

$$\frac{\dot{w}}{w} = (1 - w^2)\left(\frac{\dot{a}}{a} + \frac{\mathbf{x}' \cdot \dot{\mathbf{x}}'}{\mathbf{x}'^2}\right). \tag{5.101}$$

Alternatively, one can substitute Eq. (5.101) into (5.100) to obtain

$$\frac{\dot{\varepsilon}}{\varepsilon} + \frac{\dot{a}}{a}\dot{\mathbf{x}}^2(1 + w^2) = (1 - w^2)\frac{\mathbf{x}' \cdot \dot{\mathbf{x}}'}{\mathbf{x}'^2}. \tag{5.102}$$

Now, the total energy of a piece of string is

$$E = a \int \varepsilon U d\sigma = \mu_0 a \int \frac{\varepsilon}{w} d\sigma; \tag{5.103}$$

this trivially corresponds to a total energy density ρ. Now, part of this is the bare energy that can be ascribed to the string itself,

$$E_0 = \mu_0 a \int \varepsilon d\sigma \tag{5.104}$$

(whence one can define the bare string energy density ρ_0) while the rest is in the small-scale wiggles.

$$E_w = \mu_0 a \int \frac{1 - w}{w} \varepsilon d\sigma. \tag{5.105}$$

Each of these can be used to yield a characteristic length scale for the string network: the total length could be the length that a Goto–Nambu string with the same total energy would have, while the bare length measures the actual length. Thus we will correspondingly define a characteristic lengthscale L associated with the total energy E and a correlation length ξ associated with the bare string energy E_0. From the point of view of an analytic model, the key consequence of the existence of more than one length scale is that we are no longer allowed to identify the three natural length scales we considered in the Goto–Nambu case, namely a characteristic (energy) length scale L, the string correlation length ξ and the string curvature radius R. In other words, *we can no longer have a one-scale model*.

One also needs to rethink the way in which averages are defined. Specifically, when one is defining average quantities over the string network (say, the average

RMS velocity), should the average be over the total energy or just the energy in string

$$\langle \dot{\mathbf{x}}^2 \rangle = \frac{\int \dot{\mathbf{x}}^2 U \varepsilon d\sigma}{\int U \varepsilon d\sigma}, \qquad \langle \dot{\mathbf{x}}^2 \rangle_0 = \frac{\int \dot{\mathbf{x}}^2 \varepsilon d\sigma}{\int \varepsilon d\sigma}; \qquad (5.106)$$

in other words, should pieces of string that have larger mass currents be given more weight in the average? Given the discussion so far, it should be intuitively clear that the first definition is the natural one, but the opposite choice deserves further discussion. These two averaging procedures can be applied to any other relevant quantity. For a generic quantity Q, they are related via

$$\langle Q \rangle = \frac{\langle QU \rangle_0}{\langle U \rangle_0}. \qquad (5.107)$$

An averaged model for wiggly cosmic string evolution should contain three (rather that two) evolution equations. Apart from evolution equations for a length scale and velocity, there will be a third equation which describes the evolution of small-scale structure. This is reminiscent of the three-scale model [8], but actually there are two crucial differences. First, in the three scale model all three evolution equations do in fact describe length scales, while in our case only one of them does so (although a second equation can dependently be converted into one that does). Second, in the three scale model there is no allowance for the evolution of the string velocities. From a physical point of view, the natural way to include small-scale structure in this type of analytic model is through an evolution equation for the renormalized string mass per unit length μ, defined in the obvious way

$$\mu = \frac{E}{E_0}. \qquad (5.108)$$

5.2.1 Averaged Evolution and Energy Transfers

The averaging procedure for the transonic elastic model is in principle identical to the one followed for the Goto–Nambu case, although the added complexity will manifest itself in several ways [12]. In accordance with the above discussion, we will define averaged quantities attributing more weight to regions with more small-scale structure. Hence we take the average of a generic quantity Q to be

$$\langle Q \rangle = \frac{\int Q \frac{\varepsilon}{w} d\sigma}{\int \frac{\varepsilon}{w} d\sigma}. \qquad (5.109)$$

In particular, we will deal with the average RMS string velocity, $v^2 = \langle \dot{\mathbf{x}}^2 \rangle$ and also with the renormalised string mass per unit length $\mu \equiv E/E_0 = \langle w \rangle^{-1} = \langle w^{-1} \rangle_0$. Strictly speaking, μ is a scale-dependent quantity, $\mu = \mu(\ell, t)$ as measured in

Goto–Nambu simulations [7], but the μ thus defined is to be understood as a quantity measured at a mesoscopic scale somewhat smaller than the horizon. Intuitively, an obvious choice will therefore be the coherence length in the analytic model itself.

By differentiating the above microscopic equations, one finds the corresponding averaged evolution equations. Specifically, the total length of a given piece of string— or the corresponding density—evolves according to

$$\frac{\dot{E}}{E} = \frac{\dot{\rho}}{\rho} + 3\frac{\dot{a}}{a} = \frac{\dot{E_0}}{E_0} + \frac{\dot{\mu}}{\mu} = \left[\langle w^2 \rangle - v^2 - \langle w^2 \dot{\mathbf{x}}^2 \rangle \right] \frac{\dot{a}}{a}. \tag{5.110}$$

We similarly obtain, for the energy density in string

$$\frac{\dot{E_0}}{E_0} = \frac{\dot{\rho_0}}{\rho_0} + 3\frac{\dot{a}}{a} = \left[1 - \mu \langle w(1 + w^2)\dot{\mathbf{x}}^2 \rangle \right] \frac{\dot{a}}{a} - \frac{a\mu}{R} \langle w(1 - w^2)(\dot{\mathbf{x}} \cdot \hat{\mathbf{u}}) \rangle, \tag{5.111}$$

where R is the string curvature radius. As one would expect the Hubble expansion essentially acts on the string length, not the total length: stretching has the effect of decreasing wiggliness, just as it decreases velocity. On the other hand, curvature tends to accelerate the strings, thereby decreasing the string energy and hence tending to increase wiggliness. The evolution equation for μ is not independent, and can be obtained from the above. However, one must be careful about the choice of lengthscale at which one is defining it. If this is a fixed scale, then all one has to do is take the difference of the dynamical equations for E and E_0. *However, if we want to define it at the scale of the correlation length we must allow for the fact that this scale also evolves with time.* Generically

$$\frac{\dot{\mu}}{\mu} = \frac{\dot{E}}{E} - \frac{\dot{E_0}}{E_0} + \frac{1}{\mu}\frac{\partial \mu}{\partial \ell}\dot{\ell}, \tag{5.112}$$

where ℓ is the mesoscopic lengthscale at which we define μ. The energy terms can now be obtained from the above equations, while the scale drift term can be related to the multifractal dimension [14], to yield, ignoring terms of second order in ℓ/R,

$$\frac{\dot{\mu}}{\mu} = \frac{a\mu}{R} \langle w(1 - w^2)(\dot{\mathbf{x}} \cdot \hat{\mathbf{u}}) \rangle + \frac{\dot{a}}{a} \left[\langle w^2 \rangle - 1 + \langle (\mu w - 1)(1 + w^2)\dot{\mathbf{x}}^2 \rangle \right] + [d_m(\ell) - 1]\frac{\dot{\ell}}{\ell}. \tag{5.113}$$

This also means that an opposite drift term must be included in the equation for the density in string

$$\frac{\dot{E_0}}{E_0} = \frac{\dot{\rho_0}}{\rho_0} + 3\frac{\dot{a}}{a} = \left[1 - \mu \langle w(1 + w^2)\dot{\mathbf{x}}^2 \rangle \right] \frac{\dot{a}}{a} - \frac{a\mu}{R} \langle w(1 - w^2)(\dot{\mathbf{x}} \cdot \hat{\mathbf{u}}) \rangle - [d_m(\ell) - 1]\frac{\dot{\ell}}{\ell}, \tag{5.114}$$

Notice that this makes physical sense: E is the total energy if the network, and thus an invariant quantity, but the string energy E_0 and μ do depend on the scale at which we have decided to measure them. Finally the evolution equation for the string RMS velocity is

$$\left(\dot{v^2}\right) = \frac{2a}{R}\langle w^2(1 - \dot{\mathbf{x}}^2)(\dot{\mathbf{x}} \cdot \hat{\mathbf{u}})\rangle - \frac{\dot{a}}{a}\langle(v^2 + \dot{\mathbf{x}}^2)(1 + w^2)(1 - \dot{\mathbf{x}}^2)\rangle, \qquad (5.115)$$

and for analogous reasons there is also a scale drift term in this equation which has the form

$$\frac{\partial v^2}{\partial \ell}\frac{d\ell}{dt} = \frac{1 - v^2}{1 + \langle w^2\rangle}\frac{\partial \langle w^2\rangle}{\partial \ell}\frac{d\ell}{dt}. \qquad (5.116)$$

The physical interpretation of this term is not as simple as that for the renormalized mass, but one immediate consequence of the presence of this drift term is that strictly speaking this is no longer a purely 'microscopic' RMS velocity, but rather a 'mesoscopic' one.

The coefficient in the drift term is unity for a Brownian network ($d_m = 2$) and vanishes for straight segments ($d_m = 1$): a straight line is a straight line regardless of the scale at which one is looking at it. Naturally it also vanishes if we're considering a time-independent scale. The analogies between the evolution equations for μ and v are manifest. Observe, however, an expected but crucial difference: the curvature term in the wiggliness equation vanishes both in the tensionless and the Goto–Nambu limits, while that in the velocity equation only vanishes in the tensionless limit.

Recall that v and μ are averaged quantities; they have been put inside average signs, respectively in Eqs. (5.113) and (5.115) simply as a means to yield simpler algebraic expressions; when expanding those expressions they can be freely taken out of the averages since they have no spatial dependence. Moreover, note that in order to obtain the terms involving the curvature radius R in the above equations one needs to make use of the following identities

$$\frac{1}{\varepsilon(1 - \dot{\mathbf{x}}^2)}\left(\frac{\mathbf{x}'}{\varepsilon}\right)' \cdot \dot{\mathbf{x}} = -\frac{\mathbf{x}' \cdot \dot{\mathbf{x}}'}{\mathbf{x}'^2} = \frac{a}{R}(\dot{\mathbf{x}} \cdot \hat{\mathbf{u}}), \qquad (5.117)$$

where $\hat{\mathbf{u}}$ is a unit vector defined as

$$\frac{a}{R}\hat{\mathbf{u}} = \frac{d^2\mathbf{x}}{ds^2} \qquad (5.118)$$

$$ds = |\mathbf{x}'|d\sigma = \sqrt{1 - \dot{\mathbf{x}}^2}\varepsilon d\sigma. \qquad (5.119)$$

We also need to discuss what phenomenological terms should be added to the above equations to account for energy losses into loops and for the energy transfer between the bare string and the wiggles. Firstly we need the string correlation length, which clearly needs to be defined with respect to the bare (as opposed to the total) string density

$$\rho_0 \equiv \frac{\mu_0}{\xi^2} .$$ (5.120)

We assume that the correlation length thus defined is approximately equal to the string curvature radius defined in Eq. (5.118) and appearing in Eqs. (5.114)–(5.113), $\xi \sim R$; such an assumption can be tested numerically, and although not exact it is sufficiently accurate for our present purposes. The correlation length ξ still has a physically clear meaning, while the other characteristic length scale L is now only a proxy for the total energy in the network, $\rho \equiv \mu_0/L^2$. In analogy with what was done for the simple Goto–Nambu case, we define the fraction of the bare energy density lost into loops per unit time as

$$\left(\frac{1}{\rho_0} \frac{d\rho_0}{dt} \right)_{\text{loops}} = -cf_0(\mu) \frac{v}{\xi} .$$ (5.121)

Numerical simulations suggest that small-scale structure enhances loop production, and we allow for this possible enhancement via an explicit dependence on μ, encoded in a function $f_0(\mu)$ which should approach unity in the Goto–Nambu limit.

Importantly, in the wiggly case we have an additional phenomenological term. Whenever two strings inter-commute, kinks are produced (whether or not loop production occurs). This leads to energy being transferred from the bare string to the small-scale wiggles. We will model this as follows

$$\left(\frac{1}{\rho_0} \frac{d\rho_0}{dt} \right)_{\text{wiggles}} = -cs(\mu) \frac{v}{\xi} ,$$ (5.122)

in analogy with the above term for losses into loops. Beyond the fact that the phenomenological parameter s vanishes in the Goto–Nambu limit, its precise behavior is less obvious than the former one. Note that in particular it should include the effects of kink decay on long strings (notably due to gravitational radiation), a process that is not accounted for in numerical simulations of string networks. As for the fraction of the total energy lost into loops, we need to take into account that the energy may be come from the bare string or from the wiggles

$$\left(\frac{1}{\rho} \frac{d\rho}{dt} \right)_{\text{loops}} = \left(\frac{1}{\rho} \frac{d\rho_0}{dt} \right)_{\text{loops}} + \left(\frac{1}{\rho} \frac{d\rho_w}{dt} \right)_{\text{loops}} .$$ (5.123)

The energy loss from the bare string has already been characterized by the parameter f_0 in Eq. (5.121); defining an analogous term for the losses from the wiggles

$$\left(\frac{1}{\rho_w} \frac{d\rho_w}{dt} \right)_{\text{loops}} = -cf_1(\mu) \frac{v}{\xi} ,$$ (5.124)

we end up with

$$\left(\frac{1}{\rho}\frac{d\rho}{dt}\right)_{\text{loops}} = -c\left[\frac{f_0}{\mu} + f_1\left(1 - \frac{1}{\mu}\right)\right]\frac{v}{\xi} = -cf(\mu)\frac{v}{\xi}, \qquad (5.125)$$

where we defined an overall loss parameter f. We might expect this term to have a stronger dependence on μ than that of f_0, to account for the fact that loops are preferentially produced from regions of the long string network containing more small-scale structure than average. There is clear evidence of this fact from numerical simulations [7, 15, 16]. Somewhat similar parameters have been introduced before [8]; these are usually constant and defined as the excess kinkiness of a loop compared to a piece of long string of the same size. Here, we will explicitly include a μ dependence. As a simple illustration, if we fix $f_0 = 1$, specify that both energy loss terms strictly match the Goto–Nambu case, and recall that

$$\xi^2 = \mu L^2, \qquad (5.126)$$

we immediately get

$$f(\mu) = \sqrt{\mu}, \qquad (5.127)$$

$$f_1(\mu) = \frac{\mu^{3/2} - 1}{\mu - 1}, \qquad (5.128)$$

but again we emphasize that the detailed form of these functions warrants further discussion, and should be checked in high-resolution numerical simulations.

5.2.2 The Tensionless Limit

We now study the tensionless limit where most of the energy is in the small-scale wiggles ($w \to 0$, $T/U \to 0$), which physically corresponds to the local string tension being negligible when compared to the energy density, and hence to very high wiggliness, $\mu \gg 1$. This is in fact a very simple limit to study (for example the scale drift term in the velocity equation is negligible), but it will provide insights into the behavior of wiggles that will be very useful for future studies.

We start with the RMS velocity equation (Eq. 5.115), which yields

$$v \propto a^{-1}; \qquad (5.129)$$

and using this result in the equation for the total energy (Eq. 5.110) we have

$$E = \text{const.}, \quad \rho \propto a^{-3}. \qquad (5.130)$$

What is happening is clear. The network is conformally stretched, and since the stretching acts on the string length the wiggliness will in fact have to be decreasing. Since the network is frozen it will eventually dominate the energy density of the universe. Solving the Friedmann equation with the string density, we find

$$a \propto t^{2/3} , \tag{5.131}$$

so this string-dominated universe is like a matter-dominated universe. This physical interpretation can be confirmed by looking at the remaining equations. On a fixed scale we find

$$E_0 \propto a , \quad \rho_0 \propto a^{-2} . \tag{5.132}$$

It then follows that the network's correlation length is proportional to the scale factor

$$\xi \propto a , \tag{5.133}$$

as indeed is the case for the total length in string

$$\ell_0 = \frac{E_0}{\mu_0} = \frac{E}{U} \propto a , \tag{5.134}$$

which confirms the conformal stretching. As for the renormalized mass per unit length, since $\mu E_0 = E = \text{const.}$ we immediately have in this fixed scale case

$$\mu \propto a^{-1} ; \tag{5.135}$$

recall the mesoscopic interpretation of μ: since the network is frozen but stretched by the expansion, the effective mass per unit length on a given scale is correspondingly reduced.

In the general case where the scale where μ is defined is allowed to vary, we have

$$\frac{\dot{E}_0}{E_0} = -\frac{\dot{\mu}}{\mu} = \frac{\dot{a}}{a} - [d_m(\ell) - 1]\frac{\dot{\ell}}{\ell} . \tag{5.136}$$

If we follow a scale that is proportional to the scale factor ($\ell \propto a$), we have

$$\frac{\dot{\mu}}{\mu} = [d_m(\ell) - 2]\frac{\dot{a}}{a} , \tag{5.137}$$

and for a Brownian network μ and E_0 would both be constant. On the other hand μ will decrease (increase) for a network with a smaller (larger) fractal dimension, and E_0 will have the opposite behavior. For a constant multifractal dimension we can write

$$\mu \propto a^{d_m - 2} , \quad \rho_0 \propto a^{-(1+d_m)} , \tag{5.138}$$

$$\xi \propto a^{(1+d_m)/2} \, ; \tag{5.139}$$

and therefore the characteristic lengthscale L behaves as

$$L \propto a^{3/2} \propto t \, , \tag{5.140}$$

independently of the averaging scale and consistently with the behavior of E.

As an alternative, if we follow the horizon scale ($\ell \propto d_H$), for $a \propto t^{2/3}$ we have $d_H(t) = 3t$, and relating this to the scale factor we find (still with a constant fractal dimension)

$$\mu \propto a^{(3d_m-5)/2} \, , \quad \xi \propto a^{(1+3d_m)/4} \, ; \tag{5.141}$$

now μ will be a constant for a multifractal dimension $d_m = 5/3$. Interestingly, this is the fractal dimension of a self-avoiding random walk in three dimensions [14]. Conversely, if we follow the scale of the correlation length itself $\ell \propto \xi$ we find

$$\mu \propto a^{4/(3-d_m)-3} \, , \quad \xi \propto a^{2/(3-d_m)} \, , \tag{5.142}$$

and again a constant μ will correspond to a self-avoiding random walk, $d_m = 5/3$.

5.2.3 The Linear Limit

We now study linear limit ($w \to 1, \mu \to 1$) where the wiggliness is small and can be treated as a linear order perturbation to the Goto–Nambu case. This limit may also be reasonable as an approximation for comparisons with numerical simulations, which typically start out with very little or no small-scale wiggles. At the microscopic level, let's define

$$w = 1 - y \, , \tag{5.143}$$

where $y \ll 1$; macroscopically this corresponds to

$$\mu \approx 1 + \langle y \rangle \equiv 1 + Y \, , \tag{5.144}$$

where Y is similarly small and positive, and averaged quantities are now defined as

$$\langle Q \rangle = \frac{\int Q(1+y)\varepsilon d\sigma}{\int (1+y)\varepsilon d\sigma} \sim (\langle Q \rangle_0 + \langle QY \rangle_0)(1 - Y) \, , \tag{5.145}$$

which we can equivalently write

$$\langle Q \rangle \sim \langle Q \rangle_0 - Y \langle Q \rangle_0 + \langle QY \rangle_0 + \mathcal{O}(Y^2) \sim \langle Q \rangle_0 + corr_0(y, Q) \, . \tag{5.146}$$

As expected, to first approximation the two previously discussed averaging procedures are now equivalent for quantities independent of w. This also implies that the cross-terms now become trivial if we assume $\dot{\mathbf{x}}^2$ to be independent of w

$$\langle w^{\alpha_1} \dot{\mathbf{x}}^{2\alpha_2} \rangle \sim \langle w^{\alpha_1} \rangle \langle \dot{\mathbf{x}}^{2\alpha_2} \rangle \sim (1 - \alpha_1 Y) v^{2\alpha_2} , \tag{5.147}$$

since the neglected terms are all higher-order.

Linearizing the averaged evolution equations (Eqs. (5.110)–(5.113)), one finds

$$\frac{\dot{E}}{E} = \left[(1 - 2v^2) - 2Y(1 - v^2) \right] \frac{\dot{a}}{a} \tag{5.148}$$

$$\frac{\dot{E}_0}{E_0} = \left[(1 - 2v^2) + 2Yv^2 \right] \frac{\dot{a}}{a} - 2\frac{kaYv}{R} - [d_m(\ell) - 1]\frac{\dot{\ell}}{\ell} , \tag{5.149}$$

$$(\dot{v^2}) = 2v(1 - v^2) \left[\frac{ka}{R}(1 - 2Y) - 2v(1 - Y)\frac{\dot{a}}{a} - \frac{1 + 2Y}{2v}[d_m(\ell) - 1]\frac{\dot{\ell}}{\ell} \right] \tag{5.150}$$

$$\dot{Y} = 2Y\left(\frac{kav}{R} - \frac{\dot{a}}{a} \right) + [d_m(\ell) - 1]\frac{\dot{\ell}}{\ell} . \tag{5.151}$$

Finally, switching from conformal to physical time and introducing the previously discussed energy loss terms we end up with

$$2\frac{dL}{dt} = 2[1 + v^2 + Y(1 - v^2)]HL + cfv\left(1 - \frac{1}{2}Y \right) \tag{5.152}$$

$$2\frac{d\xi}{dt} = 2[1 + (1 - Y)v^2]H\xi + [2kY + c(f_0 + s)]v + \xi [d_m(\ell) - 1]\frac{1}{\ell}\frac{d\ell}{dt} \tag{5.153}$$

$$\frac{dv}{dt} = (1 - v^2)\left[\frac{k}{\xi}(1 - 2Y) - 2Hv(1 - Y) - (d_m(\ell) - 1)\frac{1 + 2Y}{2v\ell}\frac{d\ell}{dt} \right] , \tag{5.154}$$

$$\frac{dY}{dt} = [2kY + c(f_0 + s - f)]\frac{v}{\xi} - 2HY + [d_m(\ell) - 1]\frac{1}{\ell}\frac{d\ell}{dt} . \tag{5.155}$$

Recall that f_0, f and s are in principle functions of Y, while k is a function of velocity. We expect both $(f_0 + s - f)$ and $(d_m - 1)$ to be linear in Y.

A simple analysis of the scaling behavior of these evolution equations can now be carried out. It makes sense to consider a situation in which $\ell \propto t$ an we assume $d_m \sim 1 + 2Y$, which numerical simulations suggest may be a reasonable approximation [7]. If $\gamma_{GN} = L/t$ and v_{GN} are the scaling parameters in the Goto–Nambu case (refer to Chap. 2 for their definitions), then to first order in Y the new (wiggly) scaling parameters are obtained (for $c \neq 0$) by solving the algebraic system

$$\gamma_\xi \sim \frac{cfv}{2\left[1 - \lambda\left(1 + v^2 + \left(1 - v^2\right)Y\right)\right]} \tag{5.156}$$

$$\frac{v^2}{v_{GN}^2} \sim 1 + \left[\frac{c\lambda\left(2 - A - 2B\right) - 2k\left(1 - 2\lambda\right)}{2\lambda\left(k + c\right)}\right]Y \tag{5.157}$$

$$Y \sim \frac{2\left(\lambda - 1\right)\left[\left(2A + 1\right)k + \left(A + 1\right)c\right]}{\left[1 - 2A + 2D + \left(A - 2D\right)\lambda\right]k + \left[2D\left(1 - \lambda\right) - A\left(2 - \lambda\right)\right]c} . \tag{5.158}$$

Note the explicit dependence on the expansion rate ($a \propto t^\lambda$). The A, B, and D parameters come from writing

$$f_0 + s - f \sim AY \tag{5.159}$$

$$f_0 + s + f \sim 2(1 + BY) \tag{5.160}$$

$$f_0 + s \sim 1 + DY . \tag{5.161}$$

In particular, we can use the physical requirement that Y be positive to impose constraints on the linear term in the expansion of $s(Y)$. If we assume Eqs. (5.127) and (5.128) (meaning $A \sim -1/2 + D$, $B \sim 1/4 + D/2$, and $DY \sim s$), then

$$Y \sim \frac{2\left(\lambda - 1\right)\left[4kD + \left(1 + 2D\right)c\right]}{\left[4 - \left(1 + 2D\right)\lambda\right]k + \left[2 - \lambda\left(1 + 2D\right)\right]c} \tag{5.162}$$

and if $k > 0$ as expected, requiring Y to be positive is equivalent to imposing (recall that D is also expected to be positive)

$$D > \frac{(4 - \lambda)k + (2 - \lambda)c}{2\lambda(k + c)} ; \tag{5.163}$$

note that the right-hand side is always positive. Additionally, demanding that $Y < 1$ implies

$$D < \frac{4(k + c) - \lambda(k + 3c)}{2\lambda(5k + 3c) - 4(2k + c)} , \tag{5.164}$$

and surprisingly these two conditions are incompatible in both the matter and the radiation epoch. Indeed for the right-hand side of the previous equation to be positive one requires a fast enough expansion rate, namely

$$\lambda > \frac{4k + 2c}{5k + 3c},$$
(5.165)

This shows that it is not necessarily trivial to find natural energy loss parameters which enable non-trivial small-scale scaling, beyond the trivial $Y = 0$ solution. However one should interpret these results with some caution, as they may simply indicate that our assumptions regarding the energy loss terms, specifically Eqs. (5.127) and (5.128), may not be valid. Support for this possibility is provided by the fact that numerical simulations suggest small-scale scaling can happen in the linear limit [7]. Work is currently ongoing to explore the parameter space of the full model.

5.3 Towards Cosmic Superstrings

So far we have been considering the simplest defect models, and in most cases finding that a scaling solution exists. One may ask whether this result still holds for more complex models for example for the case of cosmic superstrings [17]. As we will see, the answer seems to be yes: the scaling solution is indeed quite robust and it persists in many cases, although there are certainly mechanisms that can suppress it.

We start by pointing out that explicit energy losses are not absolutely necessary for scaling in a cosmological context: the expansion itself may provide sufficient damping. For strings, scaling will occur for any expansion rate $\lambda > 1/2$, regardless of energy losses [18] for domain walls, the analogous threshold is $\lambda > 1/4$. Among other things this implies that for strings to scale in the matter era loop production not needed for scaling, as is known since the first generation of numerical simulations in the early 1990s [15, 16]. That said, note that the above paper also shows that in some string-dominated universes the correlation length grows as $L \propto a \propto t$, so this is formally a linear scaling solution, although its physical properties are somewhat different from the standard ones.

The next question has to do with the role of additional degrees of freedom on the string worldsheet. Perhaps the simplest such example is that of a charge, and we saw in this chapter that this need not stop scaling. More specifically these solutions fall into two regimes. If the expansion rate is large enough, the charge gets diluted and the standard scaling is recovered. Conversely if the expansion rate is small, the charge stays on the strings and there's no linear scaling solution: instead the correlation length grows more slowly and velocities decay. The transition between these two regimes depends on the balance between network energy losses and the expansion rate: with no energy losses, transition occurs for the matter era ($\lambda = 2/3$), but if energy losses are present this threshold will be lower, and for large enough losses it will be below radiation, meaning that the presence of a charge on the string need not have dramatic consequences.

Similarly, the presence of string junctions need not prevent scaling. This result has in fact been known (even if only in a specific class of models) for a long time [19]. It has also subsequently been obtained for other classes of models [20], and it can occur for any number of coupled scalar fields (under the assumption that all walls have the same tension). Finally, even the assumption of equal tensions can be relaxed: a hierarchy of different tensions need not prevent scaling. Again this result has been known for more than a decade, at least for one specific class of models [21]. The conditions under which it can be generalized are also under current study.

From the discussion above, it should be clear that the VOS approach to modelling realistic cosmic superstring networks is to isolate the various different physical processes that contribute to the dynamics of network, and try to understand the role of each one of them by using suitably simplified (toy) models. One can then put together this information into more accurate models and see what these imply for superstrings. A number of these physical processes are under control, in the sense that their role is well understood. Specifically, three key examples are

- The dynamical role of monopoles and junctions, as discussed in previous chapters.
- The role of extra dimensions (if applicable) and topology [22].
- The role of the intercommuting probabilities, first considered in [18] and then quantified in [23].

Another set of relevant physical process have been identified whose understanding still requires more work (which is being done, by many people). I will highlight three examples

- The role of additional degrees of freedom (such as charges and currents) needs to be understood in more detail. In these models a second lengthscale must necessarily be introduced, so these will no longer be 'one-scale' models. There are different ways to do this, which may be relevant in different contexts.
- The behavior of networks with various tension hierarchies needs to be better understood. A naive approach is to treat the network as a sum of sub-networks (one for each tension) and to add together 'one-scale' models for each of these, in a minimally coupled way. However this approach (even if it is done in a way that conserves energy, which is not always the case) has its limitations, and may miss some of the relevant physics.
- More related to the networks' observational consequences, there is the role of the non-trivial velocity correlations [7]. These exist for Goto–Nambu strings, and are expected to persist in superstring networks. Note that physically the observable effect of the presence of the junctions should be non-trivial velocity (anti)correlations on the network, and the natural way to model them is to use fractal-like tools.

Although a fully self-consistent analysis of the scaling properties of cosmic super-string networks still remains to be carried out, partial results from various groups (and our own ongoing work) do suggest that full scaling of all the components of the network can certainly occur in significant regions of parameter space. However, there are also regions of parameter space where only the lowest tension string scales (and the heaviest ones decay), and others where no component scales and the network may dominate the universe's energy budget. Further work on these issues is ongoing. Which regions of parameter space are natural (or indeed realistic) is a different issue, which is beyond the scope of this analysis.

Analytic and numerical work in the past three decades has gradually established the result that scaling attractor solutions are ubiquitous for cosmic defect networks. At least at the qualitative level, the physical reasons behind this are clear: they stem from the usual energy minimization mechanisms. Having said that, the solution is certainly not universal: in some case thee are additional physical mechanisms that naturally act to suppress it, while in others it can be suppressed at the cost of fine-tuning or otherwise violent assumptions.

References

1. A. Vilenkin, E.P.S. Shellard, *Cosmic Strings and other Topological Defects* (Cambridge University Press, Cambridge, 1994)
2. E. Witten, Nucl. Phys. B **249**, 557 (1985)
3. B. Carter, P. Peter, Phys. Rev. D **52**, 1744 (1995)
4. C.J.A.P. Martins, *Quantitative string evolution*, Ph.D. thesis, University of Cambridge (1997)
5. C.J.A.P. Martins, E.P.S. Shellard, Phys. Rev. D **57**, 7155 (1998)
6. M.F. Oliveira, A. Avgoustidis, C.J.A.P. Martins, Phys. Rev. D **85**, 083515 (2012)
7. C.J.A.P. Martins, E.P.S. Shellard, Phys. Rev. D **73**, 043515 (2006)
8. D. Austin, E.J. Copeland, T.W.B. Kibble, Phys. Rev. D **48**, 5594 (1993)
9. J. Polchinski, J.V. Rocha, Phys. Rev. D **75**, 123503 (2007)
10. B. Carter, Phys. Rev. Lett. **74**, 3098 (1995)
11. B. Carter, Brane dynamics for treatment of cosmic strings and vortons (1997). arXiv:hep-th/9705172
12. C.J.A.P. Martins, E.P.S. Shellard, J.P.P. Vieira, Phys. Rev. D **90**, 043518 (2014)
13. X. Martin, Phys. Rev. Lett. **74**, 3102 (1995)
14. H. Takayasu, *Fractals in the Physical Sciences* (Manchester University Press, Manchester, 1990)
15. D.P. Bennett, F.R. Bouchet, Phys. Rev. D **41**, 2408 (1990)
16. B. Allen, E.P.S. Shellard, Phys. Rev. Lett. **64**, 119 (1990)
17. E. Witten, Phys. Lett. B **153**, 243 (1985)
18. C.J.A.P. Martins, Phys. Rev. D **70**, 107302 (2004)
19. T. Vachaspati, A. Vilenkin, Phys. Rev. D **35**, 1131 (1987)
20. P.P. Avelino, C.J.A.P. Martins, J. Menezes, R. Menezes, J.C.R.E. Oliveira, Phys. Rev. D **78**, 103508 (2008)
21. P. McGraw, Phys. Rev. D **57**, 3317 (1998)
22. A. Avgoustidis, E.P.S. Shellard, Phys. Rev. D **71**, 123513 (2005)
23. A. Avgoustidis, E.P.S. Shellard, Phys. Rev. D **73**, 041301 (2006)

Chapter 6
Defects in Condensed Matter

Abstract We discuss friction-dominated vortex-string evolution using the VOS model. We explicitly demonstrate the relation between the high-energy physics approach and the damped and non-relativistic limits which are relevant for condensed matter physics. We also reproduce experimental results in this context and show that the vortex-string density is significantly reduced by loop production, an effect not included in the usual 'coarse-grained' approach.

6.1 Motivation

The concept of symmetry breaking plays a crucial role in modern physics, and one of its most interesting consequences is the formation of topological defects. These defects have been observed and studied in a wide variety of condensed matter contexts, including metal crystallization [1], liquid crystals [2, 3], superfluid helium [4, 5] and superconductors [6]. In models where they are allowed, defects will form whenever the rate of the phase transition is fast relative to the scale of the system size (in other words, a 'quench').

On the other hand, they are also believed to have formed in the early universe, and they can play an extremely important part in its evolution [7]. In this context, the conditions for their formation were first established by Kibble [8]—except for some subtleties in the case of the breaking of a gauge symmetry [9], they are entirely analogous.

The scaling evolution of vortex-string networks has been extensively studied analytically in both condensed matter [10] as well as in cosmological settings, but using rather different methods. This difference is perhaps understandable given the extremity of these two physical regimes, but it may not be necessary. Condensed matter descriptions tend to focus on a coarse-grained order parameter ϕ, providing a low-level picture of defect motion by estimating energy dissipation rates. On the other hand, high energy physicists take an 'idealized' one-dimensional view of string dynamics by integrating out the radial degrees of freedom (in the Higgs ϕ and other fields) to obtain a low-energy effective action—the Nambu action. The resulting relativistic equations of motion can then be averaged to describe the large-scale evolution

© The Author(s) 2016
C.J.A.P. Martins, *Defect Evolution in Cosmology and Condensed Matter*,
SpringerBriefs in Physics, DOI 10.1007/978-3-319-44553-3_6

of the string network. Naturally one should also account for energy loss mechanisms, such as loop production—something that is not done in condensed-matter contexts.

One of the virtues of the VOS model is including the effects of frictional forces [11, 12]. As such, it can be used to study string evolution at constant temperature, which is relevant in condensed matter contexts. Although it should be seen as the basis for further work, we show how the model is already predictive enough to be testable in laboratory experiments, following the discussion in [13, 14].

The usual condensed matter approach to vortex dynamics is based on a 'coarse-grained' complex scalar field ϕ. In quantum field theory, for example, one can consider the Abelian–Higgs model, which is a relativistic generalization of the Ginzburg–Landau theory of superconductivity. It is also of interest to consider the global version of this, that is the Goldstone model.

In high-energy physics it also proves to be convenient to adopt a one-dimensional view of string dynamics. In this description a string sweeps out a 2D worldsheet described by coordinates—one can be identified with the background time, t, while the other is space-like and simply labels points along the string (we will call it σ).

This 1D description is achieved by integrating over the radial modes of the vortex solution on the assumption that the scale of perturbations along the string is much larger than its width—thereby obtaining a low-energy effective action. For the case of a gauge (global) string, one thereby obtains the Nambu (Kalb–Ramond) action from the Abelian–Higgs (Goldstone) model, both discussed in Chap. 2. By varying these actions it is then straightforward to obtain the string equations of motion.

Since strings move through a background fluid, their motion is retarded by particle scattering. Vilenkin has shown [15] that this effect can be described by a frictional force per unit length that can be written

$$\mathbf{F}_f = -\frac{\mu}{\Gamma}\gamma\mathbf{v}\,, \tag{6.1}$$

where $\mathbf{v}$ is the string velocity, γ is the Lorentz factor and Γ is a constant damping coefficient, that can be written as the square of a characteristic propagation speed (which need not necessarily be the speed of light) times a 'friction timescale' τ_f, whose explicit value depends on the type of symmetry involved. For a gauge string, the main contribution comes from Aharonov-Bohm scattering [16], while in the global case it comes from Everett scattering [17]. For example, if the background fluid is a perfect gas, we have

$$\tau_f = \begin{cases} \frac{2\pi\hbar}{\beta}\frac{(k_B T_c)^2}{(k_B T)^3} & \text{Gauge} \\ \frac{2\pi\hbar}{\beta}\frac{(k_B T_c)^2}{(k_B T)^3}\ln\left(\frac{R}{\delta}\right)\ln^2(T\delta) & \text{Global} \end{cases} \tag{6.2}$$

where T is the background temperature and β is a numerical factor related to the number of particle species interacting with the string.

6.2 The VOS Model Revisited

For string motion in a flat background, the string equations of motion with the frictional force (6.1) can then be written

$$\frac{1}{c^2}\ddot{\mathbf{x}} + \left(1 - \frac{\dot{\mathbf{x}}^2}{c^2}\right)\frac{\dot{\mathbf{x}}}{\Gamma} = \frac{1}{\varepsilon}\left(\frac{\mathbf{x}'}{\varepsilon}\right)',\tag{6.3}$$

$$\dot{\varepsilon} + \frac{\dot{\mathbf{x}}^2}{\Gamma}\varepsilon = 0,\tag{6.4}$$

where the dimensionless parameter ε (which can be interpreted as a 'coordinate energy per unit length') is defined by

$$\varepsilon^2 = \frac{\mathbf{x}'^2}{1 - \dot{\mathbf{x}}^2/c^2},\tag{6.5}$$

and dots and primes respectively denote time and space derivatives. This proves to be particularly useful because dissipation is naturally incorporated in the decay of the coordinate energy density ε, while preserving the gauge conditions. Note that while this is a truly relativistic formalism, it is straightforward to obtain the non-relativistic limit that will be adequate to condensed matter contexts where the dynamics is friction-dominated. In this case, Eqs. (6.3)–(6.5) reduce to

$$\frac{\dot{\mathbf{x}}}{\Gamma} = -\frac{1}{\mathbf{x}'^4}\left[\mathbf{x}' \wedge (\mathbf{x}' \wedge \mathbf{x}'')\right],\tag{6.6}$$

and one can recognize the right-hand side as the friction force term (which is dominant in this limit), e.g. on a superfluid vortex [18]

Moreover, it has been shown [19] that a global string will behave as a superfluid vortex if it is introduced in a homogeneous background of the form

$$H_{ext}^{ijk} = \sqrt{\rho_h}\varepsilon^{ijk};\tag{6.7}$$

physically, this corresponds to giving it angular momentum. The interaction between this background and the string originates an additional force, known as the (relativistic) Magnus force, and (6.3) becomes

$$\frac{1}{c^2}\ddot{\mathbf{x}} + \left(1 - \frac{\dot{\mathbf{x}}^2}{c^2}\right)\frac{\dot{\mathbf{x}}}{\Gamma} = \frac{1}{\varepsilon}\left(\frac{\mathbf{x}'}{\varepsilon}\right)' + \Gamma'\dot{\mathbf{x}} \wedge \frac{\mathbf{x}'}{\varepsilon},\tag{6.8}$$

where $\Gamma' \propto \rho_h^{1/2}$; note that (6.4) remains unchanged. Alternatively we can re-write this equation as follows (temporarily setting $c = 1$)

$$\ddot{\mathbf{x}} + \left(2\frac{\dot{a}}{a} + \frac{a}{\ell_f}\right)\left(1 - \dot{\mathbf{x}}^2\right)\dot{\mathbf{x}} = \frac{1}{\varepsilon}\left(\frac{\mathbf{x}'}{\varepsilon}\right)' + \frac{1}{\varepsilon}\frac{\rho_h}{\mu}\dot{\mathbf{x}} \wedge \mathbf{m}, \qquad (6.9)$$

where

$$\mathbf{m} = \frac{4\pi\eta}{\sqrt{\rho_h}}\mathbf{x}' \qquad (6.10)$$

is the circulation vector; the energy Eq. (6.4) remains unchanged.

We can now proceed to average the string equations of motion to describe the large-scale evolution of the string network, as was also done in Chap. 2. Recall that the total string energy and the average RMS string velocity are

$$E = \mu\int \varepsilon d\sigma, \qquad v^2 \equiv \langle\dot{\mathbf{x}}^2\rangle = \frac{\int \dot{\mathbf{x}}^2 \varepsilon d\sigma}{\int \varepsilon d\sigma}, \qquad (6.11)$$

where the total string energy density obeys

$$\frac{d\rho}{dt} + \frac{v^2}{\Gamma}\rho = 0. \qquad (6.12)$$

We study the evolution of the long-string network assuming that it can be characterized by a single lengthscale L; for Brownian long strings, we can define the 'correlation length' L in terms of the network density as $\rho_\infty \equiv \mu/L^2$ as usual. Following Kibble [20], the rate of loop production from long-string collisions can be estimated to be

$$\left(\frac{d\rho_\infty}{dt}\right)_{\text{to loops}} = \rho_\infty \frac{v_\infty}{L}\int w\left(\frac{\ell}{L}\right)\frac{\ell}{L}\frac{d\ell}{L} \equiv \tilde{c}v_\infty \frac{\rho_\infty}{L}. \qquad (6.13)$$

where the loop 'chopping' efficiency $\tilde{c}$ is assumed to be constant. By subtracting these loop energy losses (6.13) we obtain the overall evolution equation for the characteristic lengthscale L,

$$2\frac{dL}{dt} = \frac{v_\infty^2}{\Gamma}L + \tilde{c}v_\infty. \qquad (6.14)$$

We can also study the evolution of the loop density and distribution, as already discussed in Chap. 2.

As for the evolution of the average string velocity v, recall that a non-relativistic equation is simply just Newton's law,

$$\frac{\mu}{c^2}\frac{dv}{dt} = \frac{\mu}{R} - \mu\frac{v}{\Gamma}. \qquad (6.15)$$

This merely states that curvature accelerates the strings while friction slows them down. On dimensional grounds, the force per unit length due to curvature should be μ over the curvature radius R. The form of the damping force can be found similarly.

A relativistic generalization of the velocity evolution equation (6.15) can be obtained more rigorously by differentiating (6.11):

$$\frac{1}{c^2}\frac{dv}{dt} = \left(1 - \frac{v^2}{c^2}\right)\left(\frac{k}{R} - \frac{v}{\Gamma}\right) . \tag{6.16}$$

This is exact up to second-order terms. In the curvature term, we have introduced R via the definition of the curvature radius vector,

$$\frac{\hat{\mathbf{u}}}{R} = \frac{d^2\mathbf{x}}{ds^2} , \tag{6.17}$$

where $\hat{\mathbf{u}}$ is a unit vector and s is the physical length along the string (related to the coordinate length σ by $ds = |\mathbf{x}'|d\sigma = \left(1 - \dot{\mathbf{x}}^2/c^2\right)^{1/2}\varepsilon d\sigma$). The dimensionless parameter k is defined by

$$\left\langle (1 - \frac{\dot{\mathbf{x}}^2}{c^2})(\dot{\mathbf{x}} \cdot \hat{\mathbf{u}}) \right\rangle \equiv kv\left(1 - \frac{v^2}{c^2}\right) . \tag{6.18}$$

Note that in the case of long strings, our 'one-scale' assumption implies that the curvature radius coincides with the correlation length, $R \equiv L$; on the other hand, for a loop of size ℓ we should have $\ell \approx 2\pi R$. The parameter k is 'phenomenological', and has been discussed in Chap. 2. Equations (6.14), (2.23) and (6.16) form the basis of our generalized 'one-scale' model, which we will now proceed to apply.

6.2.1 The Condensed-Matter Context

As we already noted, in this case the dynamics is always dominated by friction. This means that the 'correlation length' L should always be larger than the (constant) friction length, so we can take $k \sim 1$. Then the evolution equations can be approximated by

$$2\frac{dL}{dt} = L\frac{v^2}{\Gamma} + \tilde{c}v , \tag{6.19}$$

$$\frac{dv}{dt} = c^2\left(\frac{1}{L} - \frac{v}{\Gamma}\right) ; \tag{6.20}$$

the friction lengthscale and the string energy per unit length being respectively

$$\ell_f = \begin{cases} s & \text{Gauge} \\ s \ln\left(\frac{L}{\delta}\right) & \text{Global} \end{cases} \tag{6.21}$$

and

$$\mu = \begin{cases} T_c^2 & \text{Gauge} \\ T_c^2 \ln\left(\frac{L}{\delta}\right) & \text{Global} \end{cases} \tag{6.22}$$

where T_c is the temperature at which the strings form and s is a constant. We then find the following late-time asymptotic behavior:

$$L = \sqrt{1 + \tilde{c}} \, (\Gamma t)^{1/2} , \tag{6.23}$$

$$v = \frac{\Gamma}{L} . \tag{6.24}$$

Note that in both cases the asymptotic behavior of the long-string density is

$$\rho_\infty = \frac{(k_B T_c)^2}{(1+\tilde{c})\hbar s c^2 t} ; \tag{6.25}$$

the extra logarithmic dependencies in the global case cancel out. It should be emphasized that in condensed-matter analyses one does not consider loop formation, although there is experimental [2] and computational evidence for them. Our results show that loop formation can play an important evolutionary role.

The asymptotic ratio of the loop production and friction terms is a constant, which is precisely $\tilde{c}$—which in this way acquires a clearer physical meaning. As expected, increasing $\tilde{c}$ (or including loop losses in the first place) leads to a lower network scaling density and a smaller average velocity v; furthermore, the approach to the scaling regime is also faster.

Notice that in the gauge case the temperature only enters the scaling solution in the prefactor s. In the global case, the logarithmic dependence of the friction lengthscale gives rise to an additional logarithmic dependence of the scaling solution on temperature. One can define d via $L/\delta \equiv L/sd$; in the Goldstone model we then have

$$d^{-1}(T) = \sqrt{\frac{\lambda}{6}} \left(\frac{T_c}{T}\right)^3 \ln^2\left(\sqrt{\frac{\lambda}{6}}\frac{T}{T_c}\right) . \tag{6.26}$$

One expects an enhancement of loop production at the early stages, since the string velocity is high; correspondingly, there is a fast growth of the correlation length. Comparing with the gauge case one finds that the effect of the extra logarithmic terms is significant for a comparatively large time.

The $t^{1/2}$ scaling law for the characteristic lengthscale is a well-known result in the theory of phase ordering (that is, the growth of 'order'—as measured by some correlation length—by domain coarsening when a system is 'quenched' from a homogeneous phase into a broken-symmetry phase) with a non-conserved order parameter. In this context it is usually called the Lifshitz-Allen-Cahn [21] growth law, and it is widely supported by simulations and experiment [22] (see also Ref. [10] for a review). In the usual approach, one sets up a continuum description in terms of a coarse-grained order parameter ϕ and then assumes a *scaling hypothesis*, that is, that at late times there is a single lengthscale such that the domain structure is time-independent (in a statistical sense) when all lengths are rescaled by it. The growth law is usually derived by studying the dynamics of the defects in ϕ (see, for example, Sect. 3 in Bray's review [10]). An alternative approach [23] proceeds instead by comparing the global rate of energy change due to the energy dissipation to the local evolution of the order parameter; with the scaling hypothesis, the time-dependence of the lengthscale can be determined self-consistently.

In particular, the $L \propto t^{1/2}$ law has been experimentally confirmed for the evolution of a string network in a nematic liquid crystal (roughly speaking, a liquid made of rod-like molecules)—eg, see Ref. [2] where, as mentioned, loop formation and decay have been seen.

The $v \propto L^{-1} \ln L$ scaling law is also known in hydrodynamical contexts. Furthermore, it has been shown that it holds for superfluid vortex-rings [19] in the context of a modified Goldstone model. Hence the above result seems to indicate that a global string network at constant temperature asymptotically behaves as if it was made of loops of size L.

Finally, in the case of superfluid vortices, despite the additional Magnus force term, the evolution equations also have the form (6.19) and (6.20). In this case, however, the physical meaning of the friction lengthscale is not clear. Furthermore, it is also not clear how one can describe the effect of the Magnus force on the evolution of the network. This is a topic which warrants further work.

6.2.2 The Relativistic Regime

As a matter of completeness as well as mathematical curiosity, we now consider the evolution of a string network in flat space with a constant friction lengthscale when the initial conditions are such that the correlation length is much smaller than the friction lengthscale. Note that this is the opposite regime to the one usually observed in condensed matter.

The evolution equations will now be

$$2\frac{dL}{dt} = \frac{v^2}{\Gamma}L + \tilde{c}v,$$ (6.27)

$$\frac{dv}{dt} = (c^2 - v^2)\left(\frac{k(L)}{L} - \frac{v}{\Gamma}\right).$$ (6.28)

In the regime where $R \ll \ell_f$ the v-equation is independent of L, so its particularly easy to find the scaling regime

$$\frac{L}{\ell_f} = \left(\frac{L_o}{\ell_f} + 2\sqrt{2}\tilde{c}\right)\exp\left[\frac{c}{4\ell_f}(t - t_0)\right] - 2\sqrt{2}\tilde{c},$$ (6.29)

$$\frac{v}{c} = \frac{1}{\sqrt{2}}.$$ (6.30)

Hence L grows exponentially fast and quickly 'catches up' with ℓ_f; in other words, a network starting in the 'free' regime rapidly evolves to the usual friction-dominated regime. Note that if loop production is allowed, this fast growth of the correlation length will obviously mean that an extremely large number of loops is produced. In this case, the energy density in loops actually exceeds the energy in long strings. A word of caution is however needed here. In this case, loop reconnections onto the long string network should play an important role. However, since we still get an exponential growth if loop production is switched off ($\tilde{c} = 0$)—although the growth rate of L is obviously much larger for $\tilde{c} \neq 0$—our results should at least be qualitatively correct.

Thus the VOS model allows us to properly describe friction-dominated string dynamics, hence providing the first complete and fully quantitative study of string networks and their corresponding loop populations in condensed matter, as well as cosmological contexts. The fact that these results can be obtained in a model initially aimed at describing cosmic string evolution is, of course, a manifestation of the universality of symmetry breaking and defect formation phenomena, but it also lends weight to the validity of this approach because these cosmological models have been extensively tested numerically.

6.3 Summary: VOS in a Nutshell

As previously discussed the microscopic string equations of motion are

$$\ddot{\mathbf{x}} + \left(1 - \dot{\mathbf{x}}^2\right)\frac{\dot{\mathbf{x}}}{\ell_d} = \frac{1}{\varepsilon}\left(\frac{\mathbf{x}'}{\varepsilon}\right)' + \zeta\dot{\mathbf{x}} \wedge \frac{\mathbf{x}'}{\varepsilon}$$ (6.31)

$$\dot{\varepsilon} + \frac{\dot{\mathbf{x}}^2}{\ell_d}\varepsilon = 0.$$ (6.32)

All terms can be rigorously derived from the Nambu action, except the last one in (6.31) which comes from the Kalb–Ramond action and describes the Magnus force arising when a global string moves through a relativistic superfluid background [19] whose density is parametrized by ζ [13].

Now let us consider two limits of the equations of motion that are relevant in condensed matter systems. Firstly, if the damping term dominates, we find after some algebra (note that the Magnus force is not included, since it's not dissipative)

$$\frac{\dot{\mathbf{x}}}{\ell_d} = -\frac{1}{\mathbf{x}'^4}\left[\mathbf{x}' \wedge (\mathbf{x}' \wedge \mathbf{x}'')\right], \qquad (6.33)$$

where the right-hand side is the friction force term, e.g., on a superfluid vortex [18, 19]. Secondly, let us consider the non-relativistic limit but without damping. In this case we find

$$\dot{\mathbf{x}} = \frac{1}{\zeta\mathbf{x}'^2}\frac{\mathbf{x}' \wedge \mathbf{x}''}{\varepsilon}, \qquad (6.34)$$

which is the equation describing the frictionless motion of a vortex filament in an unbounded fluid [18]. By combining these two terms we can therefore reproduce the equation of motion obtained by Schwarz [18], which used an effective and more phenomenological 1D approach (based on a coarse-grained order parameter) to obtain the terms one by one. Note however that the Schwarz equation has further (subdominant) terms, coming from non-local and boundary contributions (which we have neglected altogether).

The VOS model has already described in detail, in particular in Chap. 2 for the case of cosmic strings. It includes a phenomenological term to account for the loss of energy from long strings by the production of loops—the 'loop chopping efficiency' parameter $\tilde{c}$. A further phenomenological term (characterized by a strength Σ and a characteristic length scale L_d) is also included to account for radiation back-reaction effects. By suitably averaging Eqs. (6.31)–(6.32) one can obtain the following evolution equations

$$2\frac{dL}{dt} = 2HL + \tilde{c}v + \frac{L}{\ell_d}v^2 + 8\Sigma v^6 \exp\left(-\frac{L}{L_d}\right), \qquad (6.35)$$

$$\frac{dv}{dt} = (1-v^2)\left(\frac{k(v)}{L} - \frac{v}{\ell_d}\right); \qquad (6.36)$$

here H is the Hubble parameter (relevant for cosmology) and k is the 'momentum parameter', given by

$$k(v) = \frac{2\sqrt{2}}{\pi}(1-v^2)(1+2\sqrt{2}v^3)\frac{1-8v^6}{1+8v^6}; \qquad (6.37)$$

its non-relativistic limit is $k_{\mathrm{nr}}(v) \sim 2\sqrt{2}/\pi$.

We can now take the condensed matter (non-relativistic) limit of the VOS model. All we need to do is set $H = 0$ and the damping length to a constant. One finds a stable attractor solution

$$L = \sqrt{1 + \tilde{c}} \, (\ell_d t)^{1/2} \, . \qquad v = \frac{\ell_d}{L} \, . \qquad (6.38)$$

Thus our Goto–Nambu based microscopic equations of motion and our averaged version of them successfully reproduce known condensed matter results, respectively the Schwarz equation and the $L \propto t^{1/2}$ scaling law. Note that our solution demonstrates the importance of loop production—although this is not usually included in theoretical or numerical analyses in the condensed matter context, it has been observed in experiments [2].

We will now further test our averaged model in the context of Abelian–Higgs field theory simulations and cosmology [14, 24]. These are a relativistic analogue of the Ginzburg–Landau theory of superconductivity. Before reviewing the comparisons between these simulations and the VOS model, it is worthwhile noting a byproduct of this work which relates to vortex tangles in condensed matter. In order to create quiescent initial conditions for string evolution in these simulations, instead of the relativistic equations, one starts solving the corresponding diffusive equations (refer to [24] for details). This is essentially equivalent to the non-relativistic evolution of magnetic flux-lines in a friction-dominated regime. Measurements of the string correlation length for the evolving network revealed a clear $L \propto t^{1/2}$ scaling, as illustrated in [14]. As well as confirming the VOS model prediction in this case (6.38), this has already been observed experimentally [2].

The relativistic evolution of the field theory string networks also clearly revealed the predicted scaling laws and, remarkably, the correlation length and velocities for all the simulations had a good asymptotic fit from the VOS model using the single parameter $\tilde{c} \approx 0.57$. This fit was universal regardless of whether the simulations were in flat space or in an expanding universe, or whether matter or radiation eras.

As for the massive radiation parameters, the simulations suggest $\Sigma = 0.5$ and $L_d = 4\pi$ which only affect the initial conditions. However, for global strings with massless radiation $L_d \to \infty$ there is a degeneracy between $\tilde{c}$ and Σ because they act in the same manner asymptotically. However, assuming the same loop chopping efficiency $\tilde{c} = 0.57$ for local and global cases, requires a much higher damping coefficient $\Sigma = 1.1$ for the latter as expected for massless radiation [24]. (These results and fits are also in reasonable agreement with other recent simulations of global strings in Ref. [25].) This excellent correspondence for both local and global strings appears to establish the validity of the two key ('localization' and 'thermodynamic') assumptions underlying the VOS model.

In the case of Goto–Nambu simulations [26–28], both friction and radiative backreaction are negligible. In the radiation and matter epochs, the best fit corresponds to a loop chopping efficiency parameter $\tilde{c} = 0.23$. We find that this value also approximately reproduces the earlier results of Bennett and Bouchet [26] and Allen and Shellard [27]. Thus fixing this parameter via the radiation era, our model correctly predicts the matter era scaling large-scale properties without any further tampering with parameters. We estimate the loop chopping efficiency to have the value $\tilde{c}_{ren} = 0.23 \pm 0.04$. We emphasize that we expect this to be a 'universal' parameter, independent of the cosmological scenario in which the string network is evolving. However, if one performs analogous simulations in flat (Minkowski) spacetime, one does find a different value, $\tilde{c}_{bare} = 0.57 \pm 0.04$, which coincides with the value we found above for field theory simulations. This is because the amount of small scale structure present in Goto and Nambu expanding runs is much larger than that in field theory and/or Minkowski space runs, and this has an influence on the large-scale features of the network described by the model. Hence the two values can be regarded as the 'renormalized' and 'bare' chopping efficiencies. This is discussed at greater length in [24]. The analysis in [14] shows that these parameters can we also reproduce the transition between the radiation and matter eras, and a more recent work shows that the same happens in the case of the analogous models for domain walls [29].

Thus we conclude that the VOS model successfully reproduces the large-scale features of numerical simulations of both Goto–Nambu and field theory string networks, as well as of experiments in condensed matter physics. This quantitative correspondence provides strong evidence in support of the main assumptions on which the VOS model is based, notably string 'locality' and 'thermodynamic' averaging. Our results confirm that the dominant mechanisms affecting string network evolution are loop production and damping, whether from friction or radiation depending on the context. In condensed matter systems one expects that the two are of comparable magnitude, despite loop production being neglected in the usual treatments. For global strings, loop production is dominant but radiative damping can significantly affect the network density. The outstanding issue raised by comparisons of gauged string networks in Nambu and field theory simulations remains the modelling of small-scale features on which more work (some of it ongoing) is clearly needed.

References

1. D. Mermin, Rev. Mod. Phys. **51**, 591 (1979)
2. I. Chuang et al., Science **251**, 1336 (1991)
3. M. Bowick et al., Science **263**, 947 (1994)

4. M. Salomaa, G. Volovik, Rev. Mod. Phys. **59**, 533 (1987); U. Parts et al., Phys. Rev. Lett. **75**, 3320 (1995)
5. W.H. Zurek, Nature **317**, 505 (1985); P.C. Hendry et al., Nature **368**, 315 (1994)
6. A. Abrikosov, Sov. Phys. JETP **5**, 1174 (1957)
7. A. Vilenkin, E.P.S. Shellard, *Cosmic Strings and other Topological Defects* (Cambridge University Press, Cambridge, 1994)
8. T.W.B. Kibble, J. Mod. Phys. **A9**, 1387 (1976)
9. M. Hindmarsh, A.-C. Davis, R.H. Brandenberger, Phys. Rev. **D49**, 1944 (1994); R.H. Brandenberger, A.-C. Davis, Phys. Lett. **B332**, 305 (1994)
10. A.J. Bray, Adv. Phys. **43**, 357 (1994)
11. C.J.A.P. Martins, E.P.S. Shellard, Phys. Rev. D **53**, 575 (1996)
12. C.J.A.P. Martins, E.P.S. Shellard, Phys. Rev. D **54**, 2535 (1996)
13. C.J.A.P. Martins, E.P.S. Shellard, Phys. Rev. B **56**, 10892 (1997)
14. C.J.A.P. Martins, J.N. Moore, E.P.S. Shellard, Phys. Rev. Lett. **92**, 251601 (2004)
15. A. Vilenkin, Phys. Rev. **D43**, 1060 (1991)
16. R. Rohm, Ph.D. thesis, Princeton University (1985); P. de Sousa Gerbert, R. Jackiw, Commun. Math. Phys. **124**, 229 (1988); M.G. Alford, F. Wilczek, Phys. Rev. Lett. **62**, 1071 (1989)
17. A.E. Everett, Phys. Rev. **D24**, 858 (1981); W.B. Perkins et al., Nucl. Phys. **B353**, 237 (1991)
18. K.W. Schwarz, Phys. Rev. **B38**, 2398 (1988)
19. R.L. Davis, E.P.S. Shellard, Phys. Rev. Lett. **63**, 2021 (1989)
20. T.W.B. Kibble, Nucl. Phys. B **252** 227 (1985); Erratum: (Nucl. Phys. B **261**, 750 (1985))
21. I.M. Lifshitz, Zh. Eksp. Teor. Fiz. **42**, 1354 (1962); S.M. Allen, J.W. Cahn, Acta Metall. **27**, 1085 (1979)
22. R.E. Blundell, A.J. Bray, Phys. Rev. **E49**, 4925 (1994); M. Mondello, N. Goldenfeld, Phys. Rev. **A45**, 657 (1992); H. Toyoki. J. Phys. Soc. Jpn. **60**, 1433 (1991)
23. A.J. Bray, A.D. Rutenberg, Phys. Rev. **E49**, R27 (1994)
24. J.N. Moore, E.P.S. Shellard, C.J.A.P. Martins, Phys. Rev. D **65**, 023503 (2002)
25. M. Yamaguchi, Phys. Rev. **D60**, 103511 (1999); M. Yamaguchi, J. Yokoyama, M. Kawasaki, Phys. Rev. **D61**, 061301 (2000)
26. D.P. Bennett, F.R. Bouchet, Phys. Rev. D **41**, 2408 (1990)
27. B. Allen, E.P.S. Shellard, Phys. Rev. Lett. **64**, 119 (1990)
28. C.J.A.P. Martins, E.P.S. Shellard, Phys. Rev. D **73**, 043515 (2006)
29. C.J.A.P. Martins, I.Y. Rybak, A. Avgoustidis, E.P.S. Shellard, Phys. Rev. D **93**, 043534 (2016)

Printed in the United States
By Bookmasters